THE NATURE OF CROPS
How we came to eat the plants we do

THE NATURE OF CROPS
How we came to eat the plants we do

John M. Warren
*The Institute of Biological Environmental and Rural
Sciences, Aberystwyth University, UK*

www.cabi.org

CABI is a trading name of CAB International

CABI	CABI
Nosworthy Way	38 Chauncy Street
Wallingford	Suite 1002
Oxfordshire OX10 8DE	Boston, MA 02111
UK	USA

Tel: +44 (0)1491 832111	Tel: +1 800 552 3083 (toll free)
Fax: +44 (0)1491 833508	E-mail: cabi-nao@cabi.org
E-mail: info@cabi.org	
Website: www.cabi.org	

A catalogue record for this book is available from the British Library, London, UK.

Library of Congress Cataloging-in-Publication Data

Warren, John, 1962- author. The nature of crops : how we came to eat the plants we do / Dr. John M. Warren, Director of Learning and Teaching in the Institute of Biological Environmental and Rural Sciences, Aberystwyth University, UK. pages cm. Includes index.
 ISBN 978-1-78064-508-7 (hardback : alk. paper) -- ISBN 978-1-78064-509-4 (pbk. : alk. paper) 1. Food crops--History. I. Title.

 SB175.W37 2015
 633--dc23

2014046530

ISBN-13: 978 1 78064 508 7 (hbk)
 978 1 78064 509 4 (pbk)

Commissioning editor: Rachel Cutts and Joris Roulleau
Assistant editor: Alexandra Lainsbury
Production editor: Shankari Wilford

Printed and bound in the UK from copy supplied by the authors by CPI Group (UK) Ltd, Croydon, CR0 4YY.

Contents

About the Author

Professor John Warren has a PhD based on a study of the sex-life of Groundsel (a weed, whose Latin name translates as, the common old man). From there he went on to, quite literally sow wild oats, at the University of Liverpool (while never fully understanding the origin of the phrase). He has been employed as a geneticist working on the international gene-bank for cacao at the University of the West Indies, Trinidad. More recently he has worked on topics as diverse as the ecological implications of exotic genes escaping into wild populations of gooseberries, why flowers wave in the breeze and what extra petals in buttercups can tell us about the age of meadows. John Warren is Director of Learning and Teaching at the Institute of Biological Environmental and Rural Sciences at Aberystwyth University, where he lectures on conservation in an agricultural context while enjoying growing and eating a wide range of fruits and vegetables.

Acknowledgements

The stories re-told in this volume have been imbibed during many enjoyable encounters, talking about science with so many enthusiastic plant biologists that they are too numerous to name individually. All these inspirational and informative friends deserve thanks. These remarkable people work across the entire rich spectrum of the plant sciences. They are dedicated to identifying, researching, conserving, and improving plant genetic diversity. While doing these important jobs, they also find time to inspire the next generation of aspiring botanists; in the hope their work will be continued into the distant future.

Particular thanks must go to Professors Noel Ellis, Will Haresign, Chris Pollock and Sid Thomas, for reading through the completed manuscript and curbing my more exuberant tendencies to embellish a storyline, by directing me towards more solid scientific foundations. I would also like to acknowledge the help of Natasha de Vere and Evan Pearson, at the National Botanic Garden of Wales for helping establish the blog on which this book is loosely based. A personal thank you goes to my family, for proof reading the first rough drafts of each chapter and to Frances Stoakley for editing the final document.

Dedication

This book is dedicated to all the teachers and plant breeders of the world, who between them feed our minds and bellies. Both these professions are frequently misunderstood and underappreciated. In reality they are overwhelmingly motivated by the altruistic desire to make the world a better place. To all of them, I say – thank you.

1

Introduction, the nature of natural

The entire raison d'être of this book is to try and ascertain why we eat so few of the plant species that are available to us on Earth. In attempting this feat the first chapter tries to establish whether our impoverished diet is a new phenomenon. The evidence suggests that our ancestral diets differed greatly between cultures and although some of these may have been more diverse than our own, many others would have been more monotonous. Throughout this book different elements of the problem are tackled by exploring crop biographies as case studies. In this first chapter this approach reveals that over the history of crop domestication, humans have successfully and repeatedly solved one of the most significant problems involved in transforming wild plants into crops, which is how to avoid being poisoned. This was achieved by a number of methods: by selecting plants that contain lower levels of toxic chemicals, by adapting our own biology to be better able to digest these new foods stuffs and finally by inventing methods of processing plant materials which make them safer to eat. These issues will re-emerge and are covered in greater depth in subsequent chapters.

Scientists at the Royal Botanic Gardens at Kew have estimated that the number of species of plants alive on the earth today is probably in excess of 400,000. Of these it is thought that many more than half of them could be considered edible to humans. It is entirely possible that we could eat an amazing 300,000 plant species. However, the reality is that we only consume a tiny fraction of what is possible. *Homo sapiens,* which is the most cosmopolitan of all species and one that thrives by virtue of being a supreme generalist, survives, by routinely eating only about 200 plant species. Amazingly more than half of the calories and the proteins that we derive from plants are provided by just three crops: maize, rice, and wheat. Given these remarkable statistics, the next time you hear a faddy child complaining that it does not want to eat its broccoli; you must inform the pernickety urchin that they are being offered one of the most appetizing, the most delectable, and the most scrummy things on a menu that lists 300,000 possible alternatives. If they think broccoli is repulsive then threaten them with something really disgusting. Ask them to imagine dinner tomorrow night chosen from the least palatable offerings on the list. The argument can be extended. Broccoli must be truly wonderful, because as a crop it has benefited from generations of selection, which have enhanced its taste qualities, palatability, nutritional value and yield. In contrast, most of the other 300,000 are still wild plants that taste, 'just as nature intended'.

The world's finest gourmet chefs are no better than the rest of us. They choose to cook using almost the same limited list of ingredients as everyone else. They too are confined by the conformities of our current highly restricted choice. Imagine if all the great artists opted to paint with less than one percent of the colours available on their palette. How stunned would the art establishment be at the avant-garde painter who was able to revolutionize our view of the world by introducing us to literally thousands of new colours? Surely they would walk away with the Turner Prize.

The animal kingdom provides us with an even more limited choice of things to eat. In the absence of seafood, the menu is effectively restricted to beef, pork, lamb and chicken. But we could argue that there is little to be gained by widening our horizons, because as everyone knows, all other meats from frogs, through to ostrich or crocodiles all taste just the same, 'a bit like chicken'. However, the 'a bit like chicken' phenomenon does not appear to apply in the world of plants. A raspberry is not one bit like a banana, an orange or an apple. More remarkably, a sprout is not even very much like cauliflower or kohlrabi, and as we shall discover later, these three are in fact all the same species. Given the vast array of different flavours and textures that could be available if we were adventurous enough to venture further down the menu, it really does demand that we ask the question - why do we limit ourselves to growing and eating just a couple of hundred plant species? Are these chosen few fruit and vegetables at the top of the menu really the only ones worth bothering with? Is everything else further down the list less appetizing than a sprout, and therefore just not worth contemplating? Even if that were true, it just makes the question even more intriguing, because, many of our current favourite crops, (the ones that we routinely eat today) were domesticated from wild ancestors that are virtually inedible. So what made our forebears set about the task of domesticating a twisted, chewy, fibrous wild root in the uncertain hope they would eventually arrive at the large, crunchy, tender, sweet, orange thing we recognize today as a carrot? Why dedicate thousands of years to this task rather than starting with a dandelion that has a much fleshier and potentially more promising root as a wild plant? Why did different groups of humans in different places and at different times, frequently decide to develop crops derived from the same plant families? Why have some of these crops spread around the world while others have remained local specialities? Even within a region, we need to ask why are so many of our chosen few crops related to each other, when other plant families are spurned? Have we always been so unadventurous in our tastes or are there good biological reasons for our conservatism? These questions are anything but trivial, because our favourite plant families are frequently highly poisonous and contain many highly toxic relatives. For example, the deadly nightshade family has given us staples such as potatoes, tomatoes and aubergines as well as the more unusual but intriguingly named Duke of Argyll's tea plant (or goji berry) all of which are stuffed full of toxic chemicals called alkaloids. There are still deeper layers of complexity to be explained, because sometimes we are attracted by oddities; by plants with very few related species, while on other occasions we clamour to consume plants with pungent odours and burning tastes. The smelly durian and hottest chilli peppers have their devotees who are prepared to pay the highest prices, and yet these delicacies revolt most uninitiated palates. Again and again we have ended up eating the most unlikely of crops while overlooking the vast majority of the potentially edible, even when they are commonplace.

My task here is not to play the role of Eve in the Garden of Eden and tempt you to eat of the forbidden fruit. But I do wish to provide you with new knowledge as together we attempt to try and answer the question - why do we eat the plants we do? This understanding will be cultivated by the telling of the wondrous, exotic and sometimes erotic tales that so often surround the origins of crop plants. However, perhaps Eve does contribute to the story in some part by warning us of the consequences of eating the forbidden. In its opening pages the bible may have reinforced the idea that only a very limited number of plants, those created by God on the third day that bear grain and those that bear fruit are deemed appropriate food for mankind.

Food as nature intended

In spite of, or because of our increased utilization of processed and packaged food, in recent years there has been an obsession with identifying products that are considered to be more natural, and by implication healthier. The language that surrounds the marketing of many food products reinforces that impression with dishes being described as, "Tasting just as nature intended", or "Produced in balance with nature". Paleo-diets that claim to be based on the human consumption of our hunter-gatherer ancestors are trumpeted as being able to reduce rates of heart disease and diabetes. Obscure wild berries are advertised as 'super-foods', packed with anti-oxidants that will reduce your risks of getting cancer. At a superficial level, such claims have the ring of good old-fashioned common sense, of brown bread and green and pleasant lands. However, with just a moment's thought, the idea that our modern diet can be described as anything like natural is exposed as superficial or even nonsensical. But these claims do demand that we ask, exactly what does it mean to describe something as being 'natural' in a human context in the twenty-first century?

The human is one of just a handful of species that has successfully colonized every continent on earth. Although over human history we have increasingly transported our familiar crops with every diaspora, our initial colonization was achieved by adapting our diet to encompass whatever was available locally. For centuries therefore, our diet must have varied widely from region to region, and to talk about 'a' paleo-diet is obviously a vast oversimplification. The paleo-diet of the Inuit of the far north was famously primarily meat and fish based, and not surprisingly would have contained little in the way of fresh fruit and vegetables. Five a day in an igloo would have been a real challenge, unless you include frozen pees! Further south, in more temperate regions, seasonality would have been all-important. Fruits and seeds in particular would be available in excess for brief periods. Gut content analysis of the 5,200 year old 'iceman' found in the Ötztal Alps in 1991 revealed that his last meal was primarily wild ibex meat. However, carbon and nitrogen isotope analysis of his hair revealed that this must have been very unusual, as his more typical fare was vegetarian and based mostly on grains and legumes. Further evidence that his diet was mostly plant is provided by the extent of abrasion present on his teeth. This seems to have been the norm in most cases and hunter-gatherers everywhere should more correctly be termed gatherer-hunters. In tropical rainforest regions seasonality was much less of an issue for gatherer-hunter peoples. Furthermore, these forests are famous for their high levels of biodiversity. As a consequence rainforests offer a potentially much more varied diet to their inhabitants. Indeed, it has been argued that one of the reasons that rainforests are so diverse is that they were effectively gardened by humans, who utilized so many of the native plants they contained. This is an attractively romantic theory, and probably wrong. It is certainly difficult to disentangle cause from effect here. Are rainforests highly diverse because humans have encouraged them to be so, to enable them to exploit a wide range of different resources, or have humans learned to utilize the vast range of plants that rainforests naturally contain? This dilemma may be unanswerable, and in a way the answer is as irrelevant as wondering which alternative is most natural, or is as futile as trying to generalize about a natural human diet.

Implicit in the search to identify the natural human diet is the assumption that this is somehow healthier than our modern diet. This view is almost as naive as the idea that there is a single natural human diet. First, we must recognize (with the exception of the very recent phenomenon of obesity) that human health and life expectancy is now greater than ever before in the history of our species. Although some of this can be attributed to advances in medical science and sanitation, we also need to thank agriculturalists for their contribution by securing a reliable supply of wholesome foods. Second, we need to consider the possibility that what may be a healthy diet in one lifestyle might be inappropriate in another. Thus, you might escape a heart attack eating pounds of whale blubber each day if your days are spent running across the ice being chased by a polar bear, but you might find the calories more difficult to burn if you are sat at an office desk staring at a computer screen. The third assumption behind the claim that a more primitive diet has health benefits is that humans have not changed along with their chosen foods. The fact that lactose intolerance is more frequent in populations with little history of consuming dairy products, and that diabetes is more common in areas with less ease of access to sugar; both suggest that this assumption is also simplistic. The same phenomenon has occurred with the digestion of starch-rich root crops. Through a process of gene duplication we are able to produce six times the amount of salivary amylase (one of the enzymes that digest starch) than do our close fruit-eating relatives, the chimpanzees.

Not only is there evidence that humans have adapted and altered genetically, driven by changes in our diet, there is also a growing body of science that suggests that our gut-flora is also dynamic and changes in response to what we consume. Although most of the evidence comes from grazing animals, it is clear that the populations of microbes that live within our intestines are really responsible for much of our digestive activity, and that these microbes change with our dietary intake. Although changes in our gut-flora might be rapid enough to enable us to adapt to seasonal changes in our food intake, they might struggle to cope with being thrown an unexpected chicken vindaloo. It could be argued that a more natural human diet in most places would have involved prolonged periods of monotony, as the same fruits or tubers dominated our meals for the period they were available. Although it does not include 'five a day' different fruit or vegetables, ironically, such a monotonous diet may have been quite healthy as it allowed plenty of opportunity for our gut-flora to adapt and become optimized to the predictable daily intake. This theory remains untested and is likely to remain so, because it might be difficult to attract interest in a diet that, although it varies greatly through the year, would indicate that breakfast, lunch and dinner would remain more or less identical for weeks on end.

Although earlier cultures may have eaten a greater diversity of plants than we do today, there is little compelling evidence that they cultivated vastly more crops than are available in our supermarkets. Thus the question remains – what causes our innate agricultural conservatism? Meanwhile, having explored what a natural human diet may have been; have we learnt anything about the plants we chose to consume and cultivate? The significance of seasonality and insecurity of food supply must have been highly important in many parts of the globe. It seems likely that the threat of starvation would have often driven our ancestors to eat things before they were ripe or well after their healthy 'consume-by' date. This survival strategy is likely to have driven an interest in obscure species that fruit during the lean periods; before or after other species. As we shall discover, an alternative strategy that humans have exploited is to domesticate crop plants that are easy and safe to store. Unfortunately, sometimes this can go very wrong.

Peanuts poisoning and the dangers of stored foods

More than any other crop, peanuts have a reputation for being potentially dangerous if they are stored in poor conditions. However, although there are very real risks associated with eating mouldy peanuts or even feeding them to garden birds, the story that alerted the public to this issue has been rather misrepresented.

Post Second World War, the British turned to the miraculous peanut as a saviour. In 1946 the newly elected British Labour government invested nearly 50 million pounds in the Tanganyika Groundnut Scheme. At that time the UK was still rationing food with cooking fats and protein being in particularly short supply. The idea was to grow peanuts across 600 square kilometres of eastern Africa. The entire scheme ended up as a complete fiasco with only a third of this area being cultivated and just 2,000 tons of peanuts being harvested. In agricultural circles it is widely believed that the failure of the Groundnut Scheme was related to the peanuts becoming contaminated with the mould *Aspergillus flavus*, which produces highly poisonous aflatoxins. In reality, the scheme failed primarily because of a series of logistical cock-ups and mismanagement. The ground selected was covered in thick vegetation that was difficult to clear. Imported heavy machinery was difficult to transport, maintain and use in eastern Africa, and floods washed transport links away. Meanwhile, angry elephants, rhinos, lions, crocodiles, bees and scorpions, plagued the 'Groundnut Army' of ex-military volunteers working on the project. When the crop was eventually planted, the clay-rich soil baked hard in the African sun and made the pre-roasted peanuts almost impossible to harvest. Finally, after being driven nuts by the nuts, the Groundnut Army resorted to planting sunflowers, which were ironically killed by the sun as a severe drought destroyed the crop. The scheme was cancelled in 1951.

Like many crops, the peanut is completely unknown as a wild plant and was probably produced by the natural hybridization of two different species of uncultivated ancestors. This event, which probably occurred thousands of years ago in Argentina or Bolivia, was followed by a fortuitous doubling of the genetic material allowing this new hybrid to produce fertile seeds and effectively creating the peanut as a new species. The oldest known peanut is dated at about five thousand years, which is a long time to be stuck down the back of a sofa!

By the time of Columbus, the peanut had conquered the whole of South and Central America, plus the Caribbean. Well before this the ancient Incas are known to have ground peanuts into a thick paste and thus have the best claim to have invented peanut butter. While grinding a peanut into mush may not seem much of an invention, it is one that has been claimed by several great Americans. George Washington Carver (the first African-American to have had his own dedicated national monument) is frequently credited with the invention of peanut butter along with 299 other things to do with a peanut. George Washington Carver was indeed an extraordinary man. Along with his mother and sister he was kidnapped from his slave master during the American Civil War. George alone survived, but was so weakened by the experience he was unable to work in the fields. Remarkably for the time, as an ex-slave he was able to gain both a school and college education and eventually became famous as an agricultural scientist. He dedicated his life to improving agricultural production in the southern states that had been damaged not only by the war but also by years of cotton production, which had left the soil impoverished. He did this by encouraging the use of crop rotations incorporating nitrogen-fixing peanuts and by promoting their consumption (hence the 300 different things to do with a peanut). His critics point out that many of these 300 things are in fact duplicates, including nearly 50 peanut based dyes, more than ten types of peanut flour and a similar number of fibreboards. Alternatively, his critics may just have been envious because they were never invited to the sorts of parties where you get to play 300 things to do with a peanut!

Carver, a very religious man, is said not to have patented his version of peanut butter because he believed all food was the gift of God and that humans should not profit from this divine generosity. Dr John Harvey Kellogg (of Cornflake fame) clearly did not share Carver's conviction and patented his own peanut meal shortly after in 1895. He started selling peanut butter making machines the year after that. Peanut butter solves the storage problem by heat treatment during roasting and by the addition of copious quantities of sugar and salt. While this may help prevent the growth of mould, it also radically changes the taste of the food, making it a whole lot more appetizing. This is a phenomenon that has been repeated on many occasions, and one that is important in determining exactly which plants we have domesticated and also continue to consume, even though we now have year-round supplies of fresh produce and more effective ways of storing food.

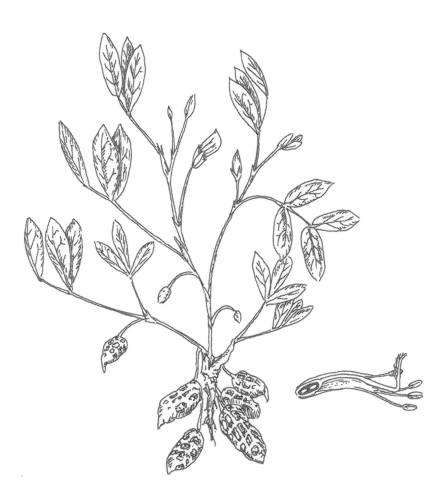

Fig. 1.1 Peanuts like many crops are prone to attack by fungi, which can result in them becoming toxic.

A corny tale of rye and the worst and best sort of food contamination

Although there may be no conclusive evidence that our natural paleo-diet was particularly healthy, perhaps the alternative does hold water. Should we worry that modern foods are so full of pesticide residues and chemical additives that they must be regarded as a health risk? The story of rye suggests that until fairly recently our diet was probably full of unexpected contaminants, and that many of these chemicals were considerably more toxic than anything we risk eating today.

For many of us rye is an unfamiliar crop, but historically it was much more frequently grown across northern Europe. Corn in comparison, has to be the most important crop of all, but confusingly the word corn refers to different crops in different places. To an American, corn is maize, to an Englishman it would be wheat, and for a Scot it could be either barley or oats. To put it simply corn is the word used in the English language to describe the most commonly grown local cereal crop (whatever that might be). Thus, if the English language was spoken in northern Europe the word corn would mean rye and similarly it would mean rice in Asia. Never mind the technicalities of species definitions, corn is the crop that people everywhere use to make the two commodities that are essential for civilized living, namely, bread and beer! Consequently, just about anywhere on the planet you will find the locals eating something that is just about recognizable as bread and drinking some form of beer. Today, this huge variety may form the basis of the beer festival or baker's window display, but historically it resulted in a story so strange that miracles and saints had to be invoked to explain it.

In those long-gone days of the Dark Ages, before western society had much scientific understanding of cause and effect, the staple grain of northern parts of Europe with its harsh cold winters and mild, damp summers was rye. In Holland, Poland, Germany and northern France where rye bread was commonly eaten, there were frequent outbreaks of what was called the Holy Fire (*ignis sacer*) or St Anthony's fire. The condition was thought to result from the sufferer being visited by the devil or being possessed by evil spirits. Those poor souls inflicted with the Holy Fire reported intense pain, burning in the skin, convulsions, and vivid hallucinations. More extreme cases were associated with the arms and legs becoming gangrenous, the limbs rapidly turning black, mummifying, drying out and most alarmingly just falling off without warning or bleeding. Outbreaks often affected entire villages. In Aquitaine, in 994 AD, 40,000 people are thought to have died from the Holy Fire. It would seem quite reasonable therefore for a traveller happening across a settlement whose inhabitants were experiencing mass hallucinations and screaming out in pain, to conclude that the population had been taken over by evil spirits, and that the only solution would be to call in the church to exorcise the entire township or burn the residents as practitioners of the dark arts.

Faced with just such a desperate situation the Bishop of Aquitaine exhibited the bones of St Martial. More famously, and as it turns out effectively, the bones of St Anthony the Hermit (not a job you see advertised much these days) were purported to offer a cure. The old saint lived originally in the desert near Alexandria. After his death his bones were taken to Constantinople when the Saracens seized Alexandria. In 1070 they were moved again by the Crusader Geslin II to Vienne in France. Here an order of monks dedicated to the shrine of St Anthony become famous for their ability to cure St Anthony's fire, and oddly enough for their abilities to perform amputations. Pilgrims to the shrine suffering from St Anthony's fire drank a liquor called St Vinage that had been exposed to the bones of the saint on Ascension Day, and it is reported that, all were cured in the space of seven days, except for those that died (which is always good to have in the small print).

In these enlightened days we are less inclined to believe in evil spirits. The true cause of St Anthony's fire and the reason that a pilgrimage to the shrine at Vienne was indeed able to affect a cure are related to the consumption of rye bread. In wet years rye is frequently infected with a fungus, which causes hard, black spike-like objects to form within the head of cereals. These ergot fungi are known to infect a wide range of grasses including most of our important cereal crops, but are especially common in rye and particularly in wetter years. Unfortunately, ergot fungi have been found to produce a range of highly toxic chemicals. Some of these have the effect of constricting blood flow and are responsible for the burning sensation in the skin and in extreme cases the loss of limbs, and because of this, old herbalists used ergots to stop the blood flow after child birth and to induce abortion. Another of the alkaloids found in ergots is closely related to lysergic acid diethylamide (LSD) and this of course is responsible for the hallucinations experienced by those suffering from St Anthony's fire. So we have a scientific explanation for the symptoms. What about the cure?

Imagine you are suffering from a burning in the skin, and your mind is playing games with you. So, on advice from the local clergy, you set off from home somewhere in northern Europe to the shrine of St Anthony in France. On foot, this takes several weeks and your supply of rye bread sandwiches would rapidly run out. The further south you travelled, the more likely it would be that the bread you ate would be made from wheat flour. Certainly it would be likely to be drier and thus less contaminated by ergots. By the time you had reached Vienne, your diet would have been free of rye and ergots long enough for your symptoms to disappear. A miraculous cure! Assuming your legs had not fallen off on route!

History is littered with tales of humans doing the weirdest of things and many have been interpreted as possible cases of ergot poisoning. For example, it has been suggested that the Salem Witch Trials in Massachusetts in 1692 and 1693 when 19 people were hung as practitioners of the dark arts were the result of them being poisoned by ergot contamination. Weather records do indicate that the two summers involved were particularly wet and that these were followed by dry summers with no records of witchcraft. As recently as 1951 in Pont-Saint-Esprit in southern France, seven people died and 50 were sent to asylums because of their erratic behaviour. This was widely reported as an outbreak of St Anthony's fire, but the true cause may have been mercury poisoning resulting from fungicide coated grain. This brings us right up to date. Warnings issued by agricultural scientists argue that eating organic food could be injurious to health because it could be potentially contaminated with toxic fungi. In contrast, organic farmers argue that these scare-stories are overstated and it is much better to consume a few natural contaminants than be poisoned by artificial chemical pesticides. Better the devil you know!

The madness of King Lear and more things to avoid eating

The story of rye suggested that our ancestors may have been naively unaware of the potential poisons that lurked within their harvested crops. However, many plant names indicate that their noxious properties were well known. For example, cow wheat (*Melampyrum pratense*) a small straggly plant that is unusual in that it grows in woodlands, heaths and formerly as an arable weed, derives its scientific name from the Greek words "melas" meaning black and "pyros" meaning wheat. This name is thought to be linked to cow wheat seeds' alleged ability to turn bread an unattractive black colour during the baking process if they contaminate flour. The archaeological excavations of the Viking settlement at York in the 1970s revealed piles of seeds in the corners of many houses. These seeds were of the rather beautiful and now very rare arable weed, corn-cockle (*Agrostemma githago*). However, historically corn-cockle was a very common plant and its seeds which are of a similar size to grains of corn were probably a contaminant in most wheat. These large and as it turns out highly toxic seeds are now easily removed by the mechanical sieving that occurs within combine harvesters. The introduction of agricultural mechanization was therefore responsible for corn-cockle's decline and thankfully for our daily bread being safe to eat. If only the Vikings of Jorvik had known this, they could have spent more time raping and pillaging rather than hand-picking the poisonous seeds out of their stores of wheat!

Further evidence that the detrimental effects of food contamination were well known in the past can be found in the works of William Shakespeare no less. In *King Lear*, Cordelia tells of seeing her mad father "Crowned with rank fumitory and furrow-weeds, with burdock, hemlock, nettles, cuckoo-flowers, darnel, and all the idle weeds that grow in our sustaining corn." Although, the exact identity of darnel is not absolutely certain, it is thought most likely to have been a weedy annual grass also known as poison darnel or cockle its Latin name being *Lolium temulentum*. The French name for darnel is ivraie, which is derived from the Latin ebriacus, meaning, intoxicated. Similarly *temulentum* also comes from Latin and means drunk. The seeds of darnel are very similar in form to grains of wheat and as with corn-cockle it was harvested along with the crop and it is thought it was a frequent contaminant of flour. As with rye, darnel itself is not toxic, but it is often associated with an endophytic fungus that grows within the plant's tissues. It is the fungus that is responsible for producing alkaloid poisons that produce the symptoms of drunkenness and may even result in death. It has been argued that not only was the madness of King Lear a result of darnel toxicity, but the practice of becoming blotto by deliberately consuming contaminated flour may have been common-place during the medieval period. In fact there are records of darnel being cultivated for its seeds which were included as an ingredient in ale recipes to enhance its intoxicating effects. Although the practice of self-inflicted darnel poisoning did involve dicing with death, the intoxicating effects were worth the risk to dull the pain of starvation and the misery of everyday life. Therefore, something that started life as a weed appears to have moved seamlessly into becoming cultivated as a drug crop.

Fig. 1.2 Darnel is an annual grass that was a common weed of arable fields that may actually have been cultivated because of its narcotic properties.

Apples of love and embracing the exotic

When trying to understand our ultra-conservatism in the limited selection of plants that we deem fit to cultivate and consume, we find a paradox. Throughout human history, so often we have been willing to rapidly assimilate the new and exotic. In some ways this is the most unusual and un-natural of all aspects of our modern diet. All other species consume the things that they find locally. In contrast, modern humans have transplanted their preferred crops into almost every patch of ground in which they would flourish. While many human immigrants experience prejudice from resident indigenous populations, the same is less likely to be true of alien crops. Thus, today wherever we travel we find familiar fruit and vegetables. Surprisingly many of these crops are the primary ingredients in national dishes and regional speciality foods. An obvious example of this is the tomato. It is virtually impossible to conceive of Italian cuisine deprived of the intensely flavoured, blood red flesh of the tomato. But of course, tomatoes are natives of Southern and Central America, and therefore, the Roman Emperors whose images adorn so many pizza palaces, could never have experienced the delights of munching on a margarita! In the UK, trying to identify any food crops that are genuinely native is an almost impossible task. The British diet and the British themselves comprise of components assembled from almost every corner of the globe. In many ways our food crops are more diverse than ever before, but at the same time we still limit ourselves to a fraction of what is edible.

Although the tomato is now routinely enjoyed around the world, this was not always the case. When it was first introduced to Europe, it was greeted with a high degree of suspicion. While not always considered good to eat, exotic plants were often believed to have incredible health giving or other properties. Like many other new foods, tomatoes were thought to be an aphrodisiac and were consequently called apples of love by the English, pomme d'amour by the French, and the devil's fruit by the Roman Catholic Church. Its Latin name *Lycopersicum* translates as wolf peach. The fruit, although eaten as a vegetable, was probably treated with suspicion because of the fact that it is a member of the same family of plants as the deadly nightshade. Indeed, unripe green tomatoes and tomato leaves do contain low levels of toxic glycoalkaloids. In addition to this, the first tomatoes introduced into Europe were probably small fruited, like modern cherry varieties, which closely resemble woody nightshade. Although not as toxic as the deadly variety, such an association would give any marketing department nightmares.

The exact date of the introduction of the tomato into Europe is uncertain. It seems probable they first arrived in Spain following Cortés' capturing of the Aztec capital city, Tenochtitlán in 1521. In spite of its toxic associations, within 23 years tomatoes were being described by the Italian botanist Mattioli as being a form of eggplant that can be cooked with salt, black pepper and oil. Even so, the majority of tomatoes grown in Italy at that time seem to have been for ornamental purposes and were used as table decorations. It took the best part of a hundred years before they became a regular component of the Italian diet. However, in comparison with the typically reserved English, this was rapid assimilation. John Gerard's famous herbal, published in 1597 considered that, "The whole plant is of a ranke and stinking savour", although he mentioned that they were eaten in Spain and other hot regions, boiled with pepper, salt and oil, "But they yeeld very little nourishment to the body, and the same naught and corrupt". The English are not known for being aficionados of foreign foods. As a result tomatoes were not widely eaten in Britain or in their American colonies for another two hundred years.

While it is easy to laugh at Gerard's lack of understanding of the merits of the tomato, most of us are equally ignorant about our foods. No other species has ever been so disconnected from the things it eats. In spite of strong regional variation in how it is prepared, around the world we tend to grow and consume many similar crops. Even in the most exotic of marketplaces, the stalls are likely to contain many easily recognizable products, along with the local specialities. Furthermore, these crops are increasingly available the year round, thanks to our ability to grow them in protected environments, to transport them rapidly and to store them in refrigerated conditions. As a result, most of us have no idea of the nationality of the crops we eat every day. Very few of us know if the plants we eat can still be found growing in the wild or are they the product of thousands of generations of cultivation and selection, and thus virtually unrecognizable from their wild progenitors? In the next chapter we will address the questions; why are there so few truly wild crops? Do these have anything in common? And if so, why would we cultivate some plants without transforming them from their wild ancestors?

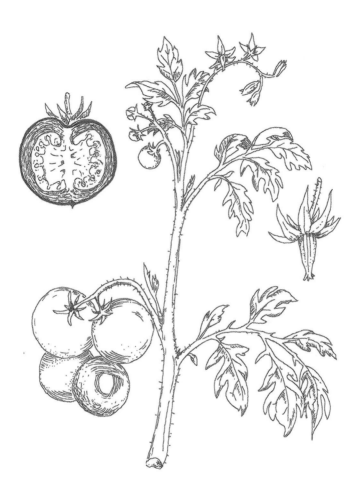

Fig. 1.3 The tomato is now loved around the world; however at first the British were reluctant to eat this crop because it has so many toxic relatives.

References

Most significant sources in the order they were utilized in this chapter.

Dickson, J.H., Oeggi, K., Holden, T.G., Handley, L.L., O'Connell, T.C. and Preston, T. (2000) The omnivorous Tyrolean Iceman: colon contents (meat, cereals, pollen, moss and whipworm) and stable isotope analyses. *Philosophical Transactions of the Royal Society of London*, 355: 1843-1849.

Moretzsohn, M.C., Hopkins, M.S., Mitchell, S.E., Kresovich, S., Valls, J.F. and Ferreira, M.E. (2004) Genetic diversity of peanut (Arachis hypogaea L.) and its wild relatives based on the analysis of hypervariable regions of the genome. *BMC Plant Biology*, 4: 11.

Zavaleta, E.G., Fernandez, B.B., Grove, M.K. and Kaye, M.D. (2001) St. Anthony's Fire (Ergotamine Induced Leg Ischemia): a case report and review of the literature. *Angiology*, 52: 349-356.

Archer, J.E., Turley, R.M. and Thomas, H. (2012) The Autumn King: remembering the land in King Lear. *Shakespeare Quarterly*, 63: 518-543.

Gerard, J. (1597) *Herball, or generall historie of plantes* [Online] Available from: <http://caliban.mpipz.mpg.de/gerarde/index.htm>. [Accessed: 22nd September 2014.]

2

Wild things

What is the difference between a crop and a wild plant? The definition is not a black and white one. In this chapter we discover that many crops hardly differ from their wild ancestors, while others have slipped from our diet and returned to the wild. Even after thousands of years of cultivation, many crops are still grown alongside wild progenitors of the same species with genes regularly flowing in both directions. Modern genetic tools are revealing that some crops have been domesticated on several occasions. In extreme cases, species that have been cultivated for millennia also occur as varieties that have just been plucked from the wild.

In the early days of the Soviet Union, the Russian scientist Nicholai Vavilov realized that most of the world's crops originated from a handful of ancient centres of domestication. Conversely, the great landmasses of Australia and North America had contributed very little to our modern diet, except for a few minor players, which are in essence still wild species, such as macadamia nuts and cranberries respectively. These Vavilovian centres of crop domestication approximately equate with early human agrarian civilizations. Thus, we can thank the Aztecs for giving us; pumpkins, maize, various beans, chocolate, papaya and many more, while the Middle East's contribution includes; apples and pears, plus many cereals such as wheat, barley, rye and oats. Globally, he identified eight such regions, which have been responsible for the domestication of almost all our crops. Vavilov recognized that these centres were not just of curious historic interest. They are of biological importance to crop geneticists such as himself.

©CAB International 2015. *The nature of crops: how we came to eat the plants we do* (J.M. Warren)

It is in areas where species first evolve or are domesticated, that we can expect to find the greatest concentration of genetic variation. This genetic diversity is the raw material, which plant breeders need to produce new crop varieties. High levels of genetic diversity are associated with Vavilovian centres, because as crops have been exported from the founder populations, genes and unusual gene combinations are lost, effectively being diluted away as sub-populations migrate from home. With this knowledge, and encouraged by the Soviet authorities Vavilov and co-workers set off around the globe to collect crop plants from each of these centres of domestication and assembled the largest collection of seeds in the world. Unfortunately, the story has a tragic end. The ideas of plant breeding based on genetic improvement did not fit comfortably with the ideology of Soviet Russia. The prevailing communist theory of the time considered that all individuals were to be considered equal. Thus, it was bit embarrassing to be searching for genetically superior genotypes, even if we are talking about potatoes. Meanwhile, Vavilov's rival, Trofim Lysenko, (who rejected the ideas of Mendelian genetics) was researching how to enhance yields by exposing seeds to cold shock treatments. In contrast, this process known as vernalization fitted beautifully the dogma of the communist regime. By changing the environment, Lysenko was able to improve productivity. This process mirrored the Stalinist vision, that by creating the perfect communist state, individuals would grow to become better, more productive members of a fairer society. Thus, it was that Vavilov's pioneering work was to become recognized in the traditional Stalinist fashion – he was sentenced to death in July 1941 a sentence that was commuted to twenty years in prison the year later and he died of malnutrition in the gulags in 1943. However, Vavilov's seed collections did manage to survive. His heroic team of scientists guarded this valuable crop genetic resource, through the Nazi siege of Leningrad from 1941-43 even though a number of them were to starve to death during the 28 month siege.

This stark rejection of the science of plant breeding was to cost the USSR dearly, and in part it contributed to the low productivity of the collective farm system and helped ensure that the Soviet State would never have the financial clout of its western rivals, where genetically superior crop varieties reigned supreme.

Bouncing cranberries

There is a degree of irony that while Soviet Russia was pioneering an understanding of the genetics of crop domestication, in the USA the quintessentially American cranberry was being cultivated for the first time. Even today the cranberry can hardly claim to be a truly domesticated crop. Its single most important variety was found growing wild in a bog in the 1840s. In contrast, its cultivation is possibly the most high-tech and strangest of any crop on earth.

The cranberry is a low growing member of the heather family, which is native to the acid peat bogs of the eastern USA and Canada. It has a close relative with twice as many chromosomes which is also called the cranberry, but which occurs in similar habitats in both Europe and North America. This European cranberry fruits only reluctantly and produces small paler berries. In parts of Scotland yet another member of the clan, the cowberry, is also known as the cranberry and has been harvested from the wild.

Long before the arrival of the Pilgrim Fathers at Plymouth Rock in 1620, the Native American Wampanoags and Narragansetts were gathering cranberries from the wild. The fruit, which they called saseminneash, was pounded with venison and fat to make pemmican. It was used in dyeing and was believed to draw the toxins from wounds caused by poison arrows – an invaluable piece of information for those family arguments over Christmas dinner that get out of hand! The European settlers quickly gained a liking for this new fruit and within 43 years 'The Pilgrims Cook Book' was describing how to make the perfect cranberry sauce; years before it became a trendy health food. At the time the berries were entirely picked from the wild. With atypical foresight for the time, the colonists and the native peoples both recognized the dangers of over harvesting. By 1670 land had been designated for the conservation of cranberries and laws regulated exactly who was allowed to exploit these public bogs. Subsequent laws forbade the picking of unripe fruit and by the 1800s many areas restricted picking to local residents.

The cranberry finally became an actively cultivated crop as the result of a chance observation made in 1816 by Captain Henry Hall, a veteran of the American War of Independence. The Captain noticed that when he inadvertently showered wild plants in a layer of sand they grew better and produced more fruit, furthermore, transplanted 'vines', which he had moved to his own farm, responded in the same way. The Cahoon family, who were pre-eminent amongst early cranberry growers, built upon this development. In 1845, Captain Alvin Cahoon built the first artificial cranberry bog which involved digging a canal to regulate the water supply. In 1847, Alvin's cousin Captain Cyrus Cahoon constructed the first level-floored cranberry bogs, some of which are still in production.

Not to be outdone, Cyrus's wife, Lettice Cahoon discovered 'Early Black' the most important cranberry variety to this day.

Modern cranberry bogs are constructed by stripping the surface vegetation to produce a level peat base. This is then covered in a layer of sand. Laser technology is used to ensure that the bog surface is absolutely level. Ditches and ponds are built to provide the water, which is an essential part of cranberry growing. To supply sufficient water for each acre of planted bog, cranberry farmers maintain a further four acres in a complex system of managed wetlands. New bogs are planted with cuttings, which take about four or five years to become productive. Each winter, between December and March, the bogs are deliberately flooded to protect the plants from frost damage. They pass the winter snug and warm below their blanket of ice. In spring a system of water sprinklers are used to spray the plants whenever a late frost is predicted.

Until the end of World War I most cranberries were still picked by hand, in spite of the fact that the first ride-on harvester was invented in 1920. Today some fruits are still dry harvested by these lawnmower like machines. However, the majority of the crop is destined for processing, and since the 1960s these berries have been harvested in a unique way. The bogs are again flooded with about 30cm of water. The fruits are then stripped from the plants by water reels, which thrash them like massive eggbeaters, and then because the berries contain pockets of air, they pop to the water's surface. The bobbing berries are then corralled by floating booms or are blown towards the banks and pumped from the bog. The finest quality, dry-harvested fruits are carefully graded by bounce testing. Soggy soft berries just don't bounce. Although this test is very simple, it is probably not something that you will want to try while purchasing a tub of cranberries from your local greengrocer!

Although cranberries remain an undomesticated crop, their cultivation is now one of the most intensive and high-tech of all. In recent years increasing demand for cranberries fuelled by marketing hype about their health benefits, has resulted in more natural bog lands being converted into cranberry farms. In Wisconsin alone cranberry beds now cover more than 15,000 acres, with a further 23,000 being used as reservoirs to supply the beds with water. Not surprisingly therefore the industry has come under pressure from the environmental lobby because of the associated loss of important wildlife habitat and because of related pesticide pollution of water ways.

My huckleberry, friend or foe

Among the select band of crops that claim North American citizenship are a number of related wild berries. As with many truly wild species, these species contain grazing deterrents, typically in the form of poisonous chemicals to defend themselves from invertebrate pests. In addition, they may also contain anti-bacterial or fungicidal compounds that have evolved to provide protection against a range of different diseases. This is a phenomenon that will be explored in more detail in chapter five. During the process of domestication, for rather obvious reasons, humans have tended to select plants with lower levels of these toxic components. In doing so, we have frequently produced crops that are not only more palatable for humans, but are also more attractive to very hungry caterpillars and are more prone to disease. Rather perversely therefore, modern agriculture is often forced to defend our crops from attack using synthetic, noxious chemical pesticides, rather than relying on the natural toxins found in ancestral crops.

Many undomesticated crops or partially domesticated crops are potentially poisonous because they contain an assortment of toxic chemicals. The fact that these chemical weapons are effective at fighting plant pathogens means that sometimes they are thought to provide miraculous medical benefits by killing the diseases that infect humans. Thus, wild crops such as cranberries, blueberries and goji berries are often given the accolade 'super-fruits'. In addition to this mechanism, reputed health-giving powers may also result from new crops containing elevated levels of anti-oxidants. These chemicals, may help prevent cancers by mopping up the highly reactive free radicals, which otherwise may react with our DNA. Typically anti-oxidants like toxic defence chemicals have been selected out during the process of domestication because they are bitter. Either way such potential properties can be influential in the naming of these wild crops.

Life can become very confusing when you start talking about some crop plants, because many are known by a large number of very different, regional names. A single name can be used time and time again to describe completely different, unrelated plants in different parts of the world. The simplest explanation for this must be human migration.

Whenever we encounter an unknown fruit or vegetable in an unfamiliar part of the world it is only natural to name it after a familiar one that it vaguely resembles from 'the old country'. Thus, Chinese gooseberries and Cape gooseberries have little in common with true gooseberries, except that they are sharp to the taste. This would be an even more convincing explanation if they actually came from China and South Africa. It may only be a partial explanation at best because many name duplications are even more completely and utterly baffling. The fig, for example, is a fruit that has been known since the dawn of time when Adam and Eve covered their embarrassment behind its hand-like, moderately sized leaves. This knowledge can make the average European male feel totally inadequate when they visit the Caribbean to find that the name fig is used to refer to the banana! It is hardly polite to enquire if the locals in these parts really need fig leaves of several metres length to hide their nether regions!

Possibly the most obvious justification for name changing occurs with widespread, undomesticated and potentially toxic crop plants that are not eaten in one part of the world, but are considered good fare in others. Here name changes can take on the function of marketing, re-branding or image enhancement. Would you fancy tucking into to a bowl of delicious poisonberries, or do wonderberries or sunberries sound a little more appetizing? I think you see the point, because poisonberries, wonderberries and sunberries are all alternative names for the fruit the British usually call black nightshade and the Americans typically call garden huckleberries.

In fact true huckleberries are a pretty difficult group of fruits to put a name to, because the term huckleberry is used for several related, low shrubby species that grow wild in the acid hill and boggy lands of Oregon, Washington and Idaho. This area is home to evergreen huckleberries, bog huckleberries, mountain huckleberries and red huckleberries. All of these species live where their name implies or look like their names suggest. The whole gang are closely related to what Americans call blueberries, which are in turn related to what the British term, bilberries, blaeberries, or whortleberries (this is not getting any simpler is it?). A few of these fruits have made it into cultivation, but generally they are not far removed from their wild ancestors and many are still actively consumed by hill walkers or even bears.

Re-branding the black nightshade as the garden huckleberry is a stroke of advertising genius, if a little economical with the truth, because the two plants are completely unrelated. The only thing they have in common is the fact that they produce small black berries, which turn a wonderful purple colour on cooking. Nightshades are related to tomatoes, potatoes and peppers, all of which contain potentially poisonous chemicals. It really is not a good idea to eat those green potatoes after all. In European culture this family has a long history of being known for its toxic properties, which has resulted in the edible species from the new world being regarded with some suspicion on their introduction. Although there are few hints in the text, Shakespeare's poisons in Hamlet, Macbeth and Romeo and Juliet are all thought to have been derived from this family of plants. The most infamous species in this family is the deadly nightshade, which has devilish associations that are lost in the darkness of medieval history. Deadly nightshade has frequently been implicated as a constituent part of witches' potions that gave them the power of flight (or at least the illusion of flying). You do not need a particularly dirty mind to realize that the image of a witch astride a broomstick has sexual connotations. Tradition has it that witches would anoint their broomsticks with ointments made from deadly nightshade before mounting them. Similarly, you do not need to be much of a biologist to realize that this method of applying the drug may speed its delivery into the blood stream, but limit the amount absorbed. The resulting hallucinations are thought to have been responsible for the sensation of flying. Although the entire ritual was associated with leaping into the air and dancing, there are no reliable records of witches actually flying.

The chemical responsible for the potentially poisonous effects of black nightshade (sorry garden huckleberries) is called solanine and this is also found in tomatoes, potatoes and green peppers. Fortunately solanine breaks down as the fruit ripen, so there are few cases of poisoning reported with any of these crops. Even so, old herbals recommend that you do not feed children black nightshade berries and that adults should only eat them when they are very ripe and preferably after a frost. The effects of solanine poisoning include excessive stimulation of the nervous system. Curiously enough atropine, which is the active toxin found in deadly nightshade causes exactly the opposite effect, as it results in depression of the nervous system. So theoretically black nightshade and deadly nightshade could be used as antidotes to each other. However, if you are feeling a bit under the weather after consuming garden huckleberries, I would not recommend following them down with a few deadly nightshade berries, unless you fancy taking your black cat for a ride across the moon on your broomstick.

Currant affairs – a little bit wild

Undomesticated crops are not restricted to North America, neither are they only found outside Vavilovian centres of domestication. However, the picture is often more complex in regions where crop plants are cultivated alongside wild populations of the same species. In these cases, it's often not clear if the wild populations are escapes from cultivation. In addition to this there could be the frequent movement of genes between farmed and non-farmed plants. In such cases feral crop plants can be regarded as semi-domesticated or just a little bit wild, as birds scatter seeds from cultivated garden plants across the countryside, and bees transport pollen between the flowers of wild and garden varieties with no prejudice about racial purity.

Until fairly recently, text books about tropical crops written by colonial agriculturists often referred to native farmers in blatantly racist language, describing 'conditions of primeval savagery almost beyond imagination'. However, these farmers were very successful as plant breeders so that it is frequently impossible to determine from which wild species their crops were originally derived. In stark contrast, of the few plant species taken into cultivation within the British Isles many are almost indistinguishable from those still growing in the wild state. Good examples of this are the red and black currants and their close relative the gooseberry. All of these can be regarded, as semi-domesticated and feral plants are commonplace but usually overlooked. The white currant is simply an albino red currant. All three of these species can be found growing wild in British hedges and wooded areas, where they differ so little from the cultivated types that some have questioned their claim to be truly native. The reason for this is two-fold: firstly, birds regularly transport fruit seeds from gardens and allotments and deposit them in the countryside. Once over the garden wall these escapees fraternized with their truly wild neighbours, to the extent that it is now often impossible to distinguish feral plants from those of mixed or wild ancestry. The second reason for wild currants and gooseberries being so similar to those found in gardens is that their domestication is surprisingly recent. Both red and black currants appear not to have been cultivated in Britain before the sixteenth century, although they must have been picked from the wild and used medicinally since at least Roman times. In fact, until the eighteenth century black currants were regarded as stinking and loathsome. In contrast gooseberry bushes were being imported from France for Edward I by the thirteenth century.

The greater antiquity of gooseberry cultivation partly explains why it is still possible to purchase in excess of 200 different gooseberry varieties, while there are less than thirty different types of red or black currant available. However, the main reason for this difference is the 'gooseberry club' craze, which swept across the north of England from Lancashire in the early nineteenth century. Although the main purpose of these clubs was to compete to produce the largest fruit, they also produced a range of different coloured fruit. The 700 plus types of gooseberry that existed at the time, not only contained green and red fruited varieties but also, yellow, white, blue and black and even striped ones. Today, although a few gooseberry clubs still survive, sadly most of these different gooseberries do not.

One of the very few ways in which domesticated currants and gooseberries differ from those growing in the wild is in their sexual preferences. In their natural condition, all three species contain incompatibility mechanisms within their flowers, which ensure that if by chance they are self-pollinated, then fertilization and seed production is prevented and the flower fails to produce a fruit. As might be expected, this tends to limit the production of fruit and thus this aversion to self-fertilization has been selected out of cultivated plants. This phenomenon has been repeated many times during the domestication of crops. The loss of their incompatibility mechanism may be responsible for reducing the barrier to crossing between species. Not only are red and black currants able to hybridize with each other and with gooseberries, but this botanical incest also extends to include other family members from the rest of Europe and North America.

Unfortunately, the transportation of currants and gooseberries across the Atlantic has had disastrous consequences. Black currants were introduced to America in 1629 and were widely grown until the 1890s. Then it was discovered that they were acting as a reservoir for a fungal disease called white-pine blister-rust, which had devastating effects on local pine trees. As a consequence of this, the growing of black currants is illegal in many parts of America. In the reverse direction, the importation of the American gooseberry was responsible for the introduction of gooseberry mildew into Britain in 1905. Although the American gooseberry is immune to this fungal disease, the leaves of European species become covered in white powder and the fruit turn into small brown blobs. The commercial cultivation of gooseberries in Britain has never recovered from this infection. The new import from America, although immune to mildew have been unable to supplant the native gooseberry, because as might be expected, they have poor taste. Modern mildew resistant varieties are therefore often hybrids between the American and European species.

There is evidence to suggest that American mildew may be responsible for the death of seedlings of native gooseberries in the UK, and thus potentially driving the extinction of wild populations. However, seedlings containing resistance genes from various North American gooseberries appear more likely to survive in the wild. This subtle Yankee pollution of the native British gene pool may be having unexpected ecological consequences. There are a number of invertebrates that have evolved to live on gooseberries and nothing else. These fussy insects seem to prefer to eat gooseberry leaves containing resistance genes because they are free from disease. However, they thrive more on genetically pure, native British plants.

Kiwifruit and kiwi berries?

It generally comes as something of a surprise to learn that the kiwifruit is an undomesticated crop. It just seems a little bit odd to think that you can still pick kiwifruit from the wild in its native China, as you might collect blackberries from the wayside in Europe. Given this, it seems even stranger to discover that the Chinese have never really eaten them, except to recommend them as a tonic for children or for women following childbirth. It was New Zealanders who first exploited the delights of this fruit, establishing the first commercial planting in 1937.

The possible reason that the Chinese failed to appreciate the potential of the kiwifruit (or yáng táo as they call it) is that many people are allergic to its fuzzy brown skin, which is covered in hairs, which act as an irritant. Not only does the kiwifruit contain allergenic factors that are commonplace in wild plants, there are in fact several different related edible species that are all eaten under the name of kiwifruit. This phenomenon is generally restricted to recently domesticated species, since species with longer histories of cultivation are more likely to be clearly differentiated from similar but undomesticated relatives.

The most commonly eaten species of kiwifruit has the readily identifiable fuzzy brown skin and is known in Latin as *Actinidia deliciosa*. The smoother skinned golden kiwifruit goes by the name *Actinidia chinensis* that gives away its Chinese, rather than New Zealand nationality. In addition, there are three very similar species that produce fruit that look like fuzzy kiwis but are about the size of grapes. These three kiwi berries are the hardy kiwi, the arctic beauty and the silver vine. None of these five species are easily distinguished from each other, because they have arisen through a process known as reticulate evolution. That means that the *Actinidia* species have a habit of hybridizing and backcrossing, so that their evolutionary family tree has branches that don't just divide, they also rejoin to form a complex lattice. This has enabled some species to evolve on more than one occasion. Kiwifruits have another very odd habit in their inheritance. Virtually all other flowering plants inherit their chloroplasts from their mothers. In kiwis, these are passed down the male line, with pollen grains carrying chloroplasts and their DNA along with the more usual nuclear DNA.

As we have already seen with huckleberries and tomatoes, newly introduced crops, like other immigrants, often change their names to avoid various forms of prejudice. It is perhaps no surprise therefore, that when the New Zealanders adopted *Actinidia deliciosa* in 1937 they should anglicize its name from the Chinese yáng táo to become called the Chinese gooseberry. Its next name change however; to the now more familiar kiwifruit, was the result of a good old-fashioned tax fiddle. In 1959 the New Zealanders started to export the fruit to the USA, as American soldiers visiting during World War II had developed a taste for them. Unfortunately, at the time, the name Chinese gooseberry or by its alternative name, the melonette both attracted higher tax duties, because berries and melons were regarded as luxury items, while fruit, which were classed as a staple, were taxed at lower rate. Plant breeders tasked with selecting for reduced hairiness in the kiwifruit claim that this is driven by pressure from female customers, who feel uncomfortable holding a pair of hirsute orbs!

Divine chocolate, wild or domesticated?

Search the world over and it is difficult to find anyone who does not like chocolate. But the sweet, smooth, melt in your mouth experience of today is a million miles from the maize thickened, chilli-flavoured, savoury drink of its origin. This is partly because most of the chocolate that we consume today can trace its ancestry through one or two generations to a tree growing wild the Amazonian rainforest. It is a truly undomesticated crop. However, other cocoa trees have a much nobler past and have been cultivated and deified for thousands of years.

Theobroma cacao, literally meaning, the food of the gods, is the Latin name of the cacao tree, the source of chocolate. It is a member of a tropical family of plants called the Sterculiaceae, which are named in honour of the Roman god of toilets, Sterculius, as they possess malodorous flowers. Cacao is a tree about the size of an apple tree, which grows wild in the gloom of the Amazonian rainforests of South America. These truly wild trees are of a kind referred to as Forastero. They are vigorous, high yielding, and disease resistant. Their pods are large, containing between 20 and 30 dark purple beans, which are surrounded by an acidic sweet sticky pulp. The beans produced from Forastero trees are known by the trade as 'bulk' and this forms the basis of milk chocolate. Deep in the upper Amazon rainforest, the sex-life of the cacao tree is most exotic. A curious physiological process within Forastero trees ensures that, if by chance, the midges which visit their flowers bring about self-pollination (fertilization with pollen from the same tree) then the tree will abort the developing pods, rather than produce inbred offspring. If cross pollinated, the pods take six months to mature and ripen, turning from green to yellow. Once ripe, the pods, which in addition to being found in the canopy, also grow directly from the trunk of the tree, are readily accessed by monkeys and squirrels, which gorge on the sweet pulp and discard the bitter tasting beans. As a curious aside, the carob tree, (well-known chocolate taste-alike) also produces its bean-like pods on its trunk, one of only two European trees to do so.

For at least 2,000 years before the Spanish conquest of the New World, the Aztec Indians of Mexico had been consuming cacao. They also used cacao beans as a form of currency, forcing their Mayan subjects to cultivate cacao to be able to pay tax. Alternatively, as few as ten cacao beans (less than a Mars bar worth) would hire the services of a whore! The custom of using cacao beans as cash apparently survived until at least 1840. Surprisingly for a 'money tree' the yield of these cultivated trees, known as Criollo is lower than those of the wild Forastero. Although, it may be possible to argue that agricultural improvement in such a crop is as unsustainable as printing bank notes, it is very strange that cacao is unique amongst crops in that the cultivated form is lower yielding than its progenitor. Criollo trees also differ in being more disease susceptible, producing small red pods with large pale violet beans. The beans are however, easier to process and considered of finer flavour. The greatest single difference between the cultivated Criollo and wild Forastero cacao is their sex-lives. Possibly related to the fact that the Mayas tended to grow cacao as solitary trees, Criollo trees indulge in self-pollination (what choice did they have). Their lower yields and higher disease susceptibility may therefore be a consequence of generations of inbreeding.

The Spanish rapidly acquired the chocolate addiction from the Indians of Central America, but preferred to add sugars rather than chilli. They established estates of the Aztec's Criollo type cacao throughout their new colonies in the Caribbean. This young industry flourished for nearly 200 years, supplying a growing demand from the cocoa houses of Europe. Then in 1725, a mystery phenomenon known as the 'blast' devastated the cacao farms throughout the entire region. Criollo cacao, which produced a fine flavoured dark chocolate, never recovered from what was probably a fungal epidemic or alternatively, but less likely, was hurricane damage. Criollo cacao in its pure form is probably now almost extinct.

Following the blast, the Caribbean rejuvenated its cacao industry by replanting with stock of the more vigorous Forastero type, imported from mainland South America. The newly imported Forastero trees then hybridized with the remnants of the old Criollo material. The populations, which arose, became known as Trinitario cacao, and are grown to this day for their bitter dark chocolate taste. As hybrids, Trinitario types are intermediate between their two parents for all characters including sexual preference. Thus, some Trinitario trees abhor inbreeding while others are content to self-pollinate.

By the time of the blast, the Spanish domination of the region was broken and the Caribbean was divided between several colonial powers and trading blocks. British islands, such as Jamaica, did not have free access to the rainforests of South America which were still held by the Spanish. Exactly how, and from where, the individual islands managed to obtain new stock to replant their estates remains a mystery. What is clear, however, is that through this quirk of colonial history the region gave rise to a wonderful range of dark chocolate varieties with exotic flavours unique to each island.

Charles Kingsley the author or "The Water-Babies" describes a more imaginative alternative version of the creation of Trinitario cacao in his book "At last: a Christmas in the West Indies". Having dismissed comets as the cause of the blast, he describes how the local Jesuit priest, Father Gumillia attributes the failure of the cacao crop to an act of God as punishment on the planters for not paying their tithes. The good priest provides support for his wrath of God theory by pointing out that a single planter, by the name of Rabelo, who duly paid his tithes was saved from the plague. However, Kingsley, being of a more rational mind set, points out that while the majority of cacao planters in Trinidad produced a fine flavoured quality product (from disease susceptible trees) that was sold well before harvesting thus craftily avoiding tax Senor Rabelo was something of a pioneer and was already growing the hardier Brazilian Forastero wild-type trees which produce lower quality beans that could only be sold after harvest and thus incurred the full tax payment. So perhaps God does not mind a bit of tax fiddling when it comes to chocolate after all.

To this day the trees, which are responsible for the distinct flavours of milk and dark chocolate, differ in their sexual habits. There is, however, one final twist to this story of sexual intrigue. Much of the world's milk chocolate is now grown in West Africa, in Ghana, Nigeria and Côte d' Ivoire. In the days of sail, when the first Old World cacao plantations were being established, seeds were imported from the nearest possible source, from trees growing along the coast of South America. These coastal populations of cacao are at the very edge of the natural range of the species and although of Forastero type, these trees are of a sub-type known as Amelonado (because of their melon like pods). The natural population density of these cacao trees is extremely low; consequently the trees have evolved the ability to self-pollinate.

The outcome of all of this historic movement of cacao trees around the world is that while potentially dark chocolate is inbred and milk chocolate more promiscuous in its habits, the reverse is more likely to be true. The cacao tree of today remains very close to its uncultivated origins and every piece of chocolate you eat could probably trace its ancestors to the forests of Amazonia in two or three generations.

So how do you get from the bitter tasting, slimy purple cacao beans to a chunk of chocolate? The first stage, surprisingly enough is fermentation. The harvested beans are piled in heaps on banana leaves or in wooden boxes. Here, attracted to the slimy pulp which surrounds the beans, fruit flies gorge themselves. The flies deliver the yeasts which start the fermentation reactions. Over several days (longer for Forastero chocolate than Criollo) the fermentation generates heat, which kills the beans and brings about a host of chemical changes, which are important in producing the complex chocolate taste. Once fermentation has finished, the beans are dried in the tropical sun before being exported. The beans are then roasted to produce the final chocolate aromas. Fifty per cent of the weight of these roasted beans is a fat called cocoa butter. This fat has the very special property that it melts in your mouth, (it also melts in other places, which is why it can also be used in suppositories). Once the cocoa butter has been extracted from the beans, the dry powder that remains is then cocoa powder. Chocolate is then made by blending many different cocoa butters and cocoa from different countries with sugar and milk solids. Good chocolate should snap rather than tear, but because most British chocolate contains more milk than does continental chocolate, it would be rather soft without the addition of other vegetable fats. This works because these fats have higher melting points than does cocoa butter, and as it happens, they are cheaper.

The origins of cacao illustrate another important aspect of the story of domestication and one that we shall return to in chapter 6. This is the importance of serendipity. There was no conscious planning behind the hybridization events that gave rise to the Trinitario varieties. Their complex pedigree is purely the product of happy, historic accidents of colonial history and piracy.

Fig. 2.1 In spite of the fact cacao has been cultivated for at least 4000 years, most of the world's chocolate is produced by effectively wild trees.

Cashew nuts and cashew apples

Without being 'food for free' fanatics, many people have experienced the delightful flavours on offer to those who glean wild blackberries, raspberries or strawberries from the countryside. With more than half of us now living in towns and cities, there are a tiny number of plant species that most people feel comfortable picking and eating from the wild. Young children are routinely warned of the danger of eating poisonous fruits. Those plants considered safe tend to be either wild berries or nuts. With these species, it is easy to see what attracted our ancestors to bring them into cultivation. In other cases however, one is left to wonder, how did anyone ever think of eating that? Such a crop is the cashew nut, which is a relative of the poison ivy, poison oak and the poison sumac.

The cashew is a tropical tree, which can grow to a height of 15 metres. It is a native of Brazil, but was already widely grown throughout South America and the Caribbean before the arrival of the Europeans. As with so many nuts, the trees are often cultivated or encouraged by humans, but they remain effectively unchanged from the wild state. The Latin name of the cashew, *Anacardium*, which means 'shaped like a heart' describes the kidney shaped nut which hangs below the bright red or yellow cashew apple (strange that). Sink your teeth into a cashew nut as it falls ripe from the tree and you quickly discover that the thick shell, which surrounds the nut, is full of toxic oil that not only tastes disgusting but also blisters your mouth. Extracting the nut without contaminating it with this noxious substance is a complex process that may involve soaking it in hot water to volatilize the oil, or burning it off by gentle roasting. Cashew yields are low, with trees being prone to more than a hundred different pests and diseases. This can result in less than thirty or forty nuts being produced per tree. No wonder they cost a packet.

It must surely have been the cashew apple, rather than its nut, which first attracted man to the tree. The cashew apple is a delicacy almost unheard of outside the tropics. It is a false fruit, which is formed from the swollen stem, which supports the cashew nut. Similar to a pear in both shape and size, the cashew apple tastes rather like a sponge soaked in an astringent sugar solution. It is no great surprise that about 95 per cent of the world's cashew apples are left to rot in the orchard. Fruit eating bats may disagree, as they are the main seed dispersal agents of the cashew, and they just love them! In Brazil, cashew apples are used to manufacture a soft drink. In the Caribbean and parts of West Africa, they are fermented to produce cashew wine, while in Goa, in India they go one better and drink cashew apple brandy. It is considered by many to taste as good a real brandy, and it certainly tastes better than the liquid extracted from the shell of the cashew nut, which is used commercially in brake linings!

The cashew has a rather interesting sex-life, in that a single tree will initially produce flowers that are entirely male, and only later in the season will it produce flowers that are hermaphrodite or occasionally female. If the male flowers are to avoid sexual frustration it requires that there is some overlap in the stages of flowering between different trees.

Fig. 2.2 The rather delicious cashew nut is protected inside a shell that is full of toxic blistering oils.

Impotent pistachios

The pistachio nut, which is a relative of the cashew, has similar issues when it comes to reproducing. The pistachio is a tree of similar stature to the cashew, but it is a native of the Old World rather than the new. Unlike the cashew, individual pistachio trees are either entirely male or entirely female. The sex-life of the pistachio is remarkable because although individual trees may be as old as 700 years, they frequently have the sexual competence of a novice. Put crudely, the problem is a simple failure to get their act to together. The most widely grown male trees, known as 'Peters', have an unfortunate habit of blossoming too soon to satisfy the female 'Kerman' trees. Other pistachio males have even greater impotency problems than do Peters, in that they are only sexually active for a couple of days each year. Peters usually manage to sustain flowering for as much as three weeks, which is sufficient in most years to allow an overlap of two frantic days with the females for wind pollination to occur. Unfortunately, all too frequently male and female pistachios fail completely to synchronize their flowering and no nuts are produced, a problem that appears particularly acute in the United States. Even when the two sexes do hit it off, pistachios only crop well every alternate year. The reason for these 'on' and 'off' years remains uncertain.

Humans have been growing and consuming pistachio nuts since antiquity. They are mentioned in the Old Testament story of Joseph. The Queen of Sheba was especially fond of them, with the assistance of a few of her court favourites, she is said to have eaten the entire pistachio crop of Assyria. Pistachio nuts remain a favourite food in the Middle East to this day, where they form an essential ingredient of wedding feasts. In spite of this ancient history, pistachios seem very little changed from the wild condition and this may not have helped its fertility problems.

Given the fact that the crop has such an ancient history of cultivation, it is remarkable that its inability to synchronize male and female flowering has not been addressed. The explanation for this may be related to female infidelity. In its native Middle East, when male pistachios fail to perform, closely related wild trees that grow in proximity to the orchards are often able to pollinate the female trees. Nut production is assured, and there is therefore little reason to select for delayed flowering males or early flowering females. When pistachio cultivation became popular in the southern United States as the result of a tax evasion scam and the banning of pistachio imports from Ayatollah Khomeini's Iran, the American female pistachios were unable to turn to their wild relatives in their times of need. When no alternative pollen source is available, the mistiming of male and female flowering inevitably results in total crop failure. Any delight felt by the Ayatollah at the impotence of the American Peters must surely have been matched by the embarrassment caused by his unfaithful Iranian females.

Cabbage so good we domesticated it twice

Ever resourceful and ingenious, mankind always seems able to find alternate uses for every plant species taken into cultivation. In many crops different varieties have been selected especially suited to their particular applications. A simple example of this phenomenon is the apple with its cooking, dessert and cider varieties. More extreme cases include: fibre and high resin forms of hemp and linseed grown for oil or flax for fibre. The undisputed master of this art is the humble cabbage. This single species occurs as spring, summer and autumn varieties, in shades of red, green and white, with Savoy or smooth round heads. But this is just the beginning, because kale, collards, kohlrabi, calabrese, broccoli, Brussels sprouts and cauliflower are still all the same species, not to mention the more obscure, ornamental kales, the palm tree cabbage and walking-stick cabbage. What is even more amazing about this long list of apparently different vegetables is that many of them appear to have been invented not just once but twice. This appears to have been relatively easy since plants, which resemble many of these different types, can still be found growing wild.

No one is entirely certain if the wild cabbage is truly native to Britain or not. Non-cultivated plants can be found growing around much of our coastline, but many of these populations are probably the descendants of plants that made it over the garden fence, rather than being relic ancient Brits. These coastal cabbages are highly variable with some plants looking more like broccoli, while in others the family resemblance is closer to kale. Whatever their origins, the cabbages that live in the wild in Britain are impressive plants. Some botanists claim that they are able to age plants by counting annual whorls of leaf scars and that individuals cabbages as old as thirty years been reported. In contrast, around the Mediterranean, plants of what is arguably the same species are unable to survive through the dry hot summers. They have adapted to the conditions by growing rapidly during the mild wetter months of winter and flowering just the once before dying. The same switch in life-style from long-lived northern types to ephemeral Mediterranean winter annuals occurs in many familiar species, including the common daisy. As we shall see later in crops such as chillies and runner beans, there are many examples of crops which are similar to the perennial cabbage in that although they have the ability to survive for many years, we tend only to grow them as annuals.

The earliest records of cultivated cabbages are to be found in the writings of ancient Greece and Rome and date from around 600BC, although they were probably grown much earlier than this. These classical cabbages must have been derived from the short-lived annual Brassicas native to the region. By the first century AD there were already accounts of varieties, which resembled leafy kales, heading cabbages, kohlrabi and something like a cauliflower or broccoli. Following the decline of the Roman Empire these early vegetables slipped into obscurity as Europe entered the dark ages.

The modern cabbage emerged in medieval Germany. Botanical genealogists have traced its ancestry to the long-lived northern forms. Unable to claim descent from the noble cabbages of ancient Rome, they have remained sour kraut ever since. As the name suggests, kohlrabi in its current incarnation is another German invention and dates from the fifteenth century. The cauliflower and its progenitor, the broccolis, are thought to have been reinvented around the Mediterranean with the island of Cyprus usually being identified as their particular Garden of Eden. This theory is supported by the fact that they are frequently fast growing annuals. However, perennial broccolis are also known, suggesting that northern Europeans have long enjoyed holiday romances in the sun.

Compared with other manifestations of the cabbage, the sprout is a new-fangled invention. It was first recorded in Belgium in 1750, near Brussels, where else? From there it took about 50 years for the crop to spread to France and Britain. This particular form of the species appears to have been completely unknown to the Emperors of Rome; the Belgians can therefore take full credit for the invention of *Brassica oleracea* var. *gemmifera*.

Recent genetic analysis has revealed that there are a number of crops, including; rice, beans, peas and bottle gourds that have all been domesticated independently on a number of occasions. However, the roman diversity of forms of cabbage remains somewhat unusual in that they seem to have returned to the wild before becoming coming back to the fold. In doing so they blur our understanding of domestication, although sometimes, it can occur instantaneously through a hybridization event, often it is best considered to be a spectrum of slow genetic change.

When the going gets tough... the tough get eaten

There are biological scientists who have dedicated entire careers to defining what a weed is; are they simply just plants in the wrong place, or do they have specific traits that make them more likely to be problematic? This debate could equally apply to the rather fuzzy definition of what is a crop, as some species, slip in and out of cultivation, depending on just how desperate we are for food at the time. These crops on the margins are sometimes referred to as 'famine crops'. In times of war, pestilence and plague humans will eat almost anything to survive. When facing dire straits the definition between what is a weed and a crop can become more blurred than ever. Wikipedia lists a whole host of interesting famine foods including cats, rats, spam even elephants. But you would imagine that you would not stay hungry for long if you had eaten the latter.

As with our beloved companion animals, in times of absolute desperation humans have resorted to eating ornamental plants. Thus, tulip, daffodils and other bulbs have been reluctantly devoured. Many of these garden exotics are likely to be toxic and probably only add to the problems of those unlucky enough to have eaten them. A less extreme alternative has been turning to fodder crops that are usually only considered fit for animals. For example in Germany the swede is a much despised crop only considered fit for livestock. However, facing extreme food shortages during both World Wars the locals were forced to eat it to survive. Meanwhile in Bonny Scotland the swede which is known locally as 'the neep' is mashed along with boiled potatoes to form part of the National dish of haggis, neeps and tatties. Fine fare indeed and anything but starvation rations.

The Scots being a hardy bunch have had their own much more extreme starvation crop. In the Outer Hebrides in times of food shortages when extreme storm conditions prevented supplies getting through from the mainland, the residents survived by eating the roots of silverweed. This scrambling plant has flowers that look rather buttercup like, but it is in fact a member of the rose family. Its tough starchy roots can be dug from sand dunes and areas of waste ground. By the time the fibrous root mat has been peeled and boiled there is very little left, so you really would need to be desperate to consider it. Even so, there are records of silverweed being dried and ground to make flour which was then used to make pancakes. The same species was also harvested by Native Americans of the northern plains, but probably only in periods when the herds of haggis had migrated south.

A more widespread reaction to avoiding starvation has been the adulteration of common food stuffs such as flour by the addition of bran, sawdust and ground tree bark. Since these additions provide little in the way of nutritional value it is probably pushing the boundaries of the definition to think of the oak tree as a crop and sawdust as a vegetable. However, the use of acorns as an alternative or adulterant to coffee is a different matter. A step back from absolute famine conditions reveals that there are many marginal crops that have acted as substitutes for more luxury items in times of restricted supply. Thus, it is not just acorns that can be been roasted to produce a coffee substitute, so have the seeds of goose-grass. This common weed which is also known as cleavers or sticky willies is actually a relative of genuine coffee. Next time the seeds of goose-grass become entangled in your socks, try peeling off their Velcro like seed coat and you will discover that although small, the pair of seeds inside are remarkable similar to coffee beans. A more commonly used alternative for coffee has been produced by roasting the roots of dandelions. However, in War time Poland the humble dandelion was cultivated as an alternative for a very different crop. The milky white latex extracted from dandelion roots was used to produce rubber. While no varieties of dandelion were grown because they yield better quality coffee, it does appear that high latex yielding types were found that produce rubber of the same quality as from the genuine rubber tree.

The ones that got away

The boundary between eating wild plants that are cultivated and foraging from the wild is a rather unclear one. There are a host of species that were previously commonly grown and eaten but for some reason they have fallen from favour and now skulk, unloved, around sites of former human habitation. In fact gardeners often complain why is it that weeds grow more vigorously than the crops they wish to tend. Revenge for being thrown over the garden fence seems an unlikely explanation. However, many of these former crops do have a habit of being pernicious weeds and perhaps that contributed to their fall from grace.

When we look at the list of species that fall into the 'former crop' group, there is an obvious simple explanation for why we have discarded them. We found a better alternative. The plants that replaced them are typically more palatable or digestible. The new improved versions contain fewer strong flavoured chemicals or tough fibres that evolved to deter herbivores. Ground elder and alexanders are two such plants. Both were probably introduced into the UK by the Romans and are a member of a widely eaten family of plants which includes; carrots, parsnips, celery and the poisonous hemlock. Both were probably domesticated in the first place because they grow early in the year and provide a spring vegetable before other alternatives were available. Unfortunately, ground elder is a second rate version of spinach and alexanders is like a pungently flavoured form of celery. In the past, such plants may have been 'forced' by growing them in dark warm conditions. This not only made them available even earlier in the season, it also tended to reduce the levels of noxious chemicals they contained and thus increased their palatability. This all seems like a lot of work, when spinach and celery are readily available. The horticultural technique of forcing has now all but died out. A last vestige remains in the forcing houses of the Yorkshire rhubarb triangle, or the upturned bucket in the garden!

It is not just minor crops that have slipped from favour. Historically, some crops that we have abandoned from our gardens appear to have previously made a significant contribution to our diet. Fat-hen, which is a member of the spinach family, could be classed as a now redundant but formerly widely eaten crop. Archaeological analysis of food remains show that fat-hen seeds were commonly mixed with cereal grains in Iron Age, Roman and Viking Europe. These seeds are higher in protein than are cereals and similar to the closely related Andean quinoa. Fat-hen has, however, not been supplanted by spinach everywhere and although it is now thought of as a weed across Europe and America, it is still cultivated for its grain, as a vegetable and for animal food in Asia and Africa.

As well as abandoning former food crops over the years, humans have also discarded a number of other crop plants as superior alternatives have been developed. Thus, the teasel, which was formerly grown for its ability to raise the nap on woollen fabrics, has now been superseded by cards of metal hooks. Evidence of its former domestication can be seen in feral populations whose seed heads are more robust and have more recurved spines than do truly wild teasels. Similarly and less well known, are the populations of stinging nettles that have escaped from cultivation. Although nettles can be consumed after boiling and are even eaten raw by a few crazy enthusiasts, they were formerly widely cultivated for the production of fabric. Thus, naturalized populations of nettles give away their former status, by being both stingless and having longer fibres in their stems that produces high quality linen-like fabric. Today nettle clothing is only available as a novelty, but formerly it was probably at least as common, as linen made from flax.

The ability of our ancestors to select out the stinging ability of nettles during the process of developing a new crop from a wild one is a neat illustration of the challenges that humans have repeatedly faced in the process of crop domestication. Such issues must surely have made some crops more attractive options for domestication than others. This is what we shall consider in the next chapter. How have humans learnt to deal with the some of the stranger habits that are found in wild plants?

Fig. 2.3 Alexanders was introduced to Britain by the Romans as an early season crop. But having been superseded by the less strongly flavoured celery it has slipped out of cultivation.

References

Most significant new sources in the order they were utilized in this chapter.

Nabhan, G.P. (2011) *Where our food comes from: retracing Nikolay Vavilov's quest to end famine*. Washington, DC: Island Press.

Smartt, J. and Simmonds, N. (1995) *The evolution of crop plants*. Second Edition. New York: Wiley-Blackwell.

Gross, P. (2010) *Superfruits*. New York: McGraw-Hill.

Warren, J. and James, P. (2006) The ecological effects of exotic disease resistance genes introgressed into British gooseberries. *Oecologia*, 147: 69-75.

Chat, J., Jáuregui, B., Petit, R.J. and Nadot, S. (2004) Reticulate evolution in kiwifruit (Actinidia, Actinidiaceae) identified by comparing their maternal and paternal phylogenies. *American Journal of Botany*, 91: 736-747.

Wood, G.A.R. and Lass, R.A. (2001) *Cocoa*. Fourth Edition. London: Wiley-Blackwell.

Vaughan, J. and Geissler, C. (2009) *The new Oxford book of food plants*. Oxford: Oxford University Press.

Darwin, C. (1868) *The variation of animals and plants under domestication* (Volume 2). London: John Murray.

Purugganan, M.D., Boyles, A.L. and Suddith, J.I. (2000) Variation and selection at the cauliflower floral homeotic gene accompanying the evolution of domesticated Brassica oleracea. *Genetics*, 155: 855-862.

Phillips, R. (2014) *Wild food: a complete guide for foragers*. London: Macmillan.

3

Learning to live with exotic sexual practices

Why are there so many species of flowering plants? The answer to this question is still hotly debated. Since the time of Darwin one of the leading theories has argued that plants have speciated by becoming genetically isolated from each other as a consequence of their individual relationships with unique species of co-evolved insects which pollinate their flowers. Insect pollination ensures that populations of plants can be genetically as separated from each other as if they were growing on different islands in the Galapagos. In this chapter we will discover that the bizarre pollination process found in many plant species has made them challenging to domesticate. This is because the necessary insect pollinators are not available widely enough or abundantly enough to allow the potential crop to be cultivated away from its home range or in sufficient quantity.

Plants so often do the most amazing things that we take them for granted. Some trees have the ability to live for thousands of years. It is entirely possible that olives are still being harvested from trees that Christ picked fruit from more than two thousand years ago. Similar things may also be true of plants that reproduce vegetatively. Although lacking the majestic grandeur of a tree, crops that regenerate from tubers may also technically live for many hundreds of years; but you would not recognize a geriatric potato if you saw one. Over evolutionary time, some species of plant may significantly increase the number of chromosomes they contain, through doubling, hybridization, or just the duplication or splitting of single chromosomes. In other plants, the amount of genetic material they

44   ©CAB International 2015. *The nature of crops: how we came to eat the plants we do*
(J.M. Warren)

contain has been reduced to the bare-minimum. Within our own species, the addition of just one extra chromosome can have dramatic negative consequences for health, life expectancy, fertility and intelligence. Plants have the ability not only to cope with such genetic variation, but also to thrive on it. Single chromosome differences can enable plants to prosper in slightly drier environments, or at marginally higher altitudes or latitudes. All of these attributes can have benefits or be problematic when it comes to making a plant a suitable candidate for domestication.

Perhaps the most remarkable things that plants do, however, are related to their highly liberal attitudes when it comes to embracing unusual sexual practices. Routinely, flowers contain mechanisms that prevent self-pollination. These can be built into the morphology of the flower, or more routinely, are physiological processes that block the germination or growth of selfed pollen. Such processes can be of great assistance to crop breeders wishing to generate hybrid plants. However, an inability to self-pollinate can often limit the production of seeds and fruit in cultivation when plants are grown in isolation or with genetically similar individuals, as we have already seen with the pistachio. Generally therefore, during the process of crop domestication, we have tended to select out promiscuity and encourage extreme incest - the ability to successfully fertilize oneself and thus elevate yield. It is just as well the church has no interest in the ethics of plant sex. The process of promoting the ability to self-pollinate has generally been relatively easy, since most mechanisms that prevent self-fertilization are not 100 per cent successful, and inbreeding individuals can be rapidly identified.

Since the time of Darwin it has been thought that the exotic sexual habits of plants may help explain why there is so much diversity within the flowering plants. The argument goes as follows: For the majority of species, to be able to evolve into two new species, it generally requires that they are genetically isolated by geographic factors, for example, finding themselves living on different islands. In these circumstances, they are not only exposed to different environments, but their inability to interbreed allows them to rapidly adapt to the new conditions. Even limited interbreeding between islands ensures that the two populations remain similar and prevents them from evolving into different species. Flowering plants have found a way around the need for geographic isolation in order to speciate. The vast majority of flowers are insect pollinated. Flowers compete for the attention of these insects. At the same time many insects survive by feeding on pollen or nectar, and thus they compete for access to flowers. Under these circumstances, flowers that become adapted to offer a reward to a unique insect species may guarantee its fidelity, and in return the insects can be assured of a secure food supply. Darwin recognized that flowers which offer nectar rewards secreted at the base of a long tube must be visited only by insects with extremely long tongues. The result of this degree of specialization is that pollen from highly modified flowers is only transported to similar flowers of the same species by their own dedicated co-adapted pollinating insect. Such flowers are so unlikely to cross-pollinate with the flowers of other species that they might as well be isolated on a different island. Thus, flowers isolated by the fidelity of their insect pollinators may be able to evolve as quickly as those isolated on remote islands. At the same time, the insects also evolve into new species adapted to be faithful to a particular floral form. This mechanism is thought to explain the explosive radiation of both flowering plants and insect species during the Cretaceous period and why these two groups of organisms contain more species than all the others combined.

Recently, however, this theory has been questioned, because many flowers have fairly simple structures and are generalists that can be pollinated by many species of insect. Similarly, many pollinating insects are also generalists and will visit and pollinate many different types of flowers. Even so, many plant families are characterized by having highly specialized flowers and these families tend to be highly diverse, containing many species. This degree of floral specialization may enable rapid speciation, but it may be an obstacle during the process of domestication.

Vanilla for one

With more than 20,000 species, the orchids are the largest family of flowering plants. About five per cent of all flowering plant species are orchids. They are highly cosmopolitan, occurring almost everywhere on earth, and yet orchids are still evocative of the rare and exotic. In temperate regions most orchids are ground dwelling perennial herbs. However the majority of orchids live in the tops of tropical trees, some are climbers and others, devoid of green pigments, live like fungi, obtaining nourishment from decaying matter. The great diversity of the orchid species is thought to result directly from the complexity of their floral forms. Darwin described much of this complexity in his book, 'On the various contrivances by which British and foreign orchids are fertilized by insects'. With so many orchids relying upon highly co-adapted and probably equally rare insects for their pollination, this must have really limited their capacity to be exploited as a crop. It must be very difficult to ensure successful pollination and seed set in orchids if you export them from their native home and separate them from their bespoke insect pollinators. Thus, in spite of the fact that there are 20,000 orchids that could be exploited as crops, very few have ever been domesticated. The vanilla orchid is often said to be the only species of direct economic importance. This of course neglects the immensely important horticultural trade in ornamental orchids, but these do not rely on producing fruit to be of value. Indeed many of these are sterile hybrids.

Today there are just three species of orchids cultivated for vanilla. The Tahitian vanilla, which is grown in French Polynesia and Hawaii, is used almost entirely in the perfume industry. A second species, the West Indian vanilla, produces Pompona vanilla, which has a fragrance reminiscent of cherries. Its active constituent, heliotropin, is used in soaps, perfumes and to flavour tobacco. It is also occasionally blended with true vanilla.

The true vanilla orchid is a native of Central America where it was used by the Aztecs of Mexico to flavour cocoa. In 1520 the Spanish Conquistador, Cortés, was given some of this drink, flavoured with chocolate and vanilla, by Montezuma. This was Europe's first taste of vanilla. Within a decade dried pods were being exported to Spain. The vanilla plant is a scrambling vine, which can grow to 15 metres tall. It has spectacular pale yellow flowers, which measure about ten centimetres across. The flowers are constructed in such a way that the male and female organs are kept apart, physically preventing self-pollination. In its native Central America, vanilla flowers are pollinated by bees and humming birds. The seedpods, which subsequently develop, are referred to as beans, although the seeds they contain, like the seeds of all orchids, are dust like, measuring less than half a millimetre across. It is these seedpods which are harvested as the source of vanilla.

Most of the world's vanilla is now grown on tropical islands such as Madagascar, Mauritius, Réunion and Tonga. However, when it was first introduced to these far-flung shores, it failed to produce any vanilla beans and was propagated entirely by cuttings. The problem was simple: far from home, vanilla was unable to pull the birds or attract the local bees. With no interest from the birds and bees, vanilla endured almost 50 years of sexual inactivity on the island of Réunion. It was not until 1841 that the former slave, Edmond Albius, managed to ease the vanilla's sexual frustration by manual means. Using a bamboo splint to replace the humming bird, a reliable method of artificially pollinating the vanilla flower was developed. Most of the world's vanilla is now produced in this fashion. This is obviously very labour-intensive and costly, but it remains economically viable purely because of the high cash values of the crop.

For other potential crop species with complex specialized flowers, manual pollination would be out of the question if the resulting fruit did not command high prices.

About eight months after pollination, the vanilla pods are ready for harvesting and curing. This is a complex process of drying in which the pods are spread in the sun for a few hours each day and then wrapped in blankets and stored in airtight containers to sweat overnight. After two weeks of this treatment, the beans turn black and are then dried off in the sun for two months. Finally the beans develop their full aroma in conditioning bins. The whole procedure takes about six months. This is why vanilla is so costly, being the second most expensive spice behind saffron.

The elaborate curing process results in dried beans that contain less than three per cent vanillin; this is a relatively simple chemical with the formula $C_8H_8O_3$ that is responsible for the vanilla taste. Synthetic vanillin was first produced from the sap of pine trees in 1874. Since then vanillin has been manufactured from lignin, a waste product of the paper industry. Five grams of artificial vanilla has approximately the same strength of flavour as a litre of natural vanilla extract and costs about one hundredth of the price to make. But then, as you pour custard over your rhubarb, it no longer evokes the same images of beautiful orchids growing on Tropical Islands.

The name orchid is derived from the Greek word for testicle; the paired root-tubers of many temperate orchids resemble the male genitals. One of the tubers, being last year's growth, is old and withered and the second, being the current year's, is fresh and full. It is no great surprise therefore to find that orchid tubers have a long association with sexual superstitions. Witches were said to use the fresh tubers in producing potions of true love, while utilizing the withered ones for spells of a more carnal nature. The herbalist Culpeper wrote that orchid roots 'provoke venery, strengthen the genital parts and help conception'. That has to rank as one of the strangest reasons for a plant to be persecuted.

Although the vanilla orchid is now the only orchid regularly consumed by man, this has not always been the case. Until the rise of the coffee shop, the streets of London were full of salopian shops, purveying salep, which was a kind of nutritious starchy soup made from the roots of various orchids. It was said to be the ideal breakfast for chimney sweeps, who were able to purchase a basin of salep in Fleet Street for three-half pence. Herbalists compared the virtues of the salep derived from our different native orchids. It is strange to think that these now rare plants were once turned into something as mundane as a bowl of soup and even stranger to discover that salep is still popular today in Turkey.

Fig. 3.1 Because of its exotic sex life, vanilla is the only orchid cultivated for food.

The birds and the beans

Trying to domesticate a crop that is pollinated by humming birds, outside of the natural range of these tiny creatures, is obviously asking for trouble. But vanilla is not the only crop that was originally pollinated by humming birds. In most parts of the world, truly red flowers (rather than shades of pink or purple) tend to be rare. The reason for this is that although insects can see into the ultra violet end of the spectrum, they have little ability to see wave lengths of light that appear red to humans. In contrast, birds are often attracted to red light. It is no surprise, therefore, that many bird-dispersed berries signal their ripeness by changing colour from cryptic green to showy red. Similarly, many flowers adapted to be pollinated by birds have bright red flowers. They tend to be more robust than insect pollinated flowers and produce considerably greater quantities of nectar.

In Central and Southern America four closely related beans were domesticated. In the cool uplands, runner beans; in warmer temperate regions, common or French beans; the Lima or butter bean was grown in sub-tropical climes and the tepary bean was cultivated in semi-arid zones. In the high Andes water boils at well below 100 degrees centigrade. To cope with this, the Incas developed beans that did not require boiling, but could be consumed after rapid frying. Although all four of these beans are now usually cultivated as annuals, actually only the tepary bean is truly so. In its native seasonally arid habitat, the tepary bean has evolved unable to survive for more than a single year. In contrast, French beans, runner beans and butter beans can all live for several years. But being intolerant of frosts this is impractical when cultivated in temperate regions.

These American beans arrived in Europe via different routes. The runner bean was introduced to Britain in the early seventeenth century by the plant collector, John Tradescant the younger, son of the famous plant collector John Tradescant the elder, in whose honour the tradescantia is named. For about 100 years, runner beans were cultivated purely as ornamentals for their attractive red flowers which resemble sweet peas, until Philip Miller of the Physic Garden in Chelsea rediscovered that the pods were good to eat.

The bright, scarlet-red flowers of runner beans are not only attractive to gardeners. In their native South America they are visited by humming birds. This fact probably partly explains why many runner bean flowers fail to produce beans when grown in European gardens and mutant white flowered varieties are considered to set more pods. Ignorant of this fact, many gardeners attribute the failure of pod production to erratic water supply. The other three cultivated American beans are self-fertile and do not require the visit of a bird or a bee to produce seeds. The flower of the broad bean is a complex trip trap, which when triggered showers visiting bees with pollen, although some modern varieties are able to produce beans in the absence of a pollinating insect. Careful inspection of a few broad bean flowers will frequently reveal that instead of fighting the trip mechanism to gain access to the flower, bumblebees often bite holes at the base of the flowers to plunder their contents.

Behind the fig leaf

There is another crop plant that has even more highly specialized flowers than the orchids and even more specialized pollinators. Their flowers are so complex that manual pollination is unfeasible, so humans have had to resort to another trick to ensure that fruit are produced in the absence of insects. This crop is the fig, which like vanilla boasts lots of close relatives. There are about 850 species of mostly tropical tree in the genus *Ficus*; all of which are edible, but only one is truly domesticated.

The fig has an association with mankind as ancient as the Garden of Eden, where Adam and Eve first used its leaves to cover their nakedness. Archaeological records show that figs have been consumed rather than used as clothing since Neolithic times.

Figs are thought to have been first cultivated in Western Asia, but spread rapidly to the Mediterranean, where they were to become a staple food in ancient Greece. The Spartans fed their athletes almost exclusively on figs, believing it would increase their strength and speed – a phenomenon now termed as having the runs! The Greeks prized their figs so much that laws were passed forbidding the export of the best fruit. The word sycophant has been linked to this legislation, as its original usage was to describe loathsome individuals who informed the authorities about fig smuggling. Sycophant literally means 'shower of figs'. The fruit was also important to the ancient Romans who regarded it as sacred because, according to legend, Romulus and Remus were suckled by a wolf in the shade of a fig tree.

Over the years the fig has not only screened the sexuality of Adam and Eve and many classical statues, it has also been successful in concealing its own. With the weirdest sex-life of any crop, it is easy to imagine why the fig may have wished to keep it habits private. Only in the twentieth century were the fig's sexual secrets finally exposed. The fig tree has been able to achieve this because of the unique structure of the fig itself.

The flowers of the fig tree are small and unisexual, being male or female or sterile-female. They occur together on mass, as do the many small flowers that comprise a dandelion head. However, in the fig, the flowers are located on the inside of a pear-like structure, which opens to the world through a small hole at the bottom. This structure is common to all species of fig and is termed the syconium.

The cultivated fig occurs in two distinct forms, a wild form or Caprifig and the edible female fig. It is impossible to tell them apart until they first flower at about seven years of age. The wild form flowers three times a year, in perfect synchrony with the life cycle of its pollinator, the fig wasp, which lives inside its syconium. The first flowering of the year occurs in early spring. The spring, or mamme, syconium, contains only male and sterile-female flowers. Within the sterile-female flowers, the larva of the fig wasp develop and pupate. By late spring the male wasps emerge first, with one thing on their mind – sex. They seek out the virgin female wasps still within their pupa, gnaw their way inside to the unsuspecting females, have sex with them, and then die. A simple and short life. Shortly afterwards, the newly fertilized female wasps emerge, just as the male fig flowers mature. Thus, as the wasps exit the syconium via the small entrance hole, they are showered with pollen. Unlike the male fig wasp, which is almost blind, legless and wingless, the females of the species are winged. They fly to the summer syconium or profichi and enter again via the small hole. This time they find only female flowers, both fertile and sterile ones. Each wasp lays about 250 single eggs within the sterile-female flowers and scatters pollen over the fertile ones. They are physically unable to insert their ovipositors into the wrong flowers because the fertile flowers are too long. The next generation of wasps develop as before, pupate, have sex and die or disperse in time to catch the final flowering of the year. The autumn syconium or mammoni, contain only sterile-female flowers and their function appears solely to allow the fig wasps to overwinter. The 'fruit' produced by the Caprifigs as a result of all this activity is leathery, resinous and inedible. All 2,000 or so different species of fig have a similar sex-life and almost all have a species of fig wasp all of their own, with both trees and wasps completely dependent on each other for their survival.

Nothing about the sex-life of the fig is simple or straightforward. There are three different kinds of fig trees that produce edible female figs. All of these produce two, rather than three crops of figs per year. In the most primitive form, the Smyrna fig, both sets of figs require pollination by fig wasps in order to produce edible figs. In the most advanced type, the common fig, all its figs are produced automatically without the need for pollination or the presence of wasps. The third form, or San Pedro fig, is intermediate between these in that its spring figs called the breba crop are produced without requiring pollination, but the main summer crop still needs the assistance of the fig wasp.

All this complexity appears to have confused fig farmers until into the twentieth century. Guides for fig growers gave hints on hanging branches of Caprifigs in female fig trees so that the pollen may be shaken out into the developing figs. Another technique recommended was to pollinate the figs by hand by inserting a small feather into each fig. Modern advice for fig farmers is based on complex calculations on the numbers of wasps expected to emerge from each Caprifig tree (between 200 and three million) and how many wasps are required to successfully pollinate each edible fig - about five!

Fig. 3.2 Figs have one of the most bizarre pollination mechanisms of any plant, but in cultivation, the fruit are usually produced automatically.

Hop, skip and a snooze

Trying to untangle the complexities of crop domestication is not only made more difficult by the strange things that plants do, it is also complicated by the many different uses that humans wish to exploit in their crop plants. Generally, we want our crops to be able to set seeds and produce fruit. Thus, often, domestication involves trying to encourage plants to have more sex. However, there are a few occasions where seed production is best avoided and the job of the agriculturalist becomes one of preventing botanical sexual activity. This is true in the banana (which will be considered later) since finding a fertilized seed in your banana would be like crunching your teeth on a marble. Another example of this is hop growing, but this is only true in the UK.

According to the old saying 'hops, reformation and beer all came to England in one year'- that of 1524. This is in fact not true, as demonstrated by the Worcestershire village of Himbleton that derives its name from the Anglo-Saxon for hop yard. The hop is indeed the only member of the Cannabis family, which is considered native to Britain, and like its more exotic relative, it is also said to have mind-altering properties. The fact that we fail to appreciate these more subtle influences is probably because they are frequently swamped by the effects of alcohol, with which hops are usually associated. However, this has not always been the case, and it is likely that hops were first cultivated for their 'herbal' properties or at least for their antiseptic abilities to prevent bacterial spoilage of food. It is highly likely that hops were first added to beer as a preservative rather than for their distinctive flavour. As we shall discover in later chapters, the ability to prevent bacterial and fungal spoilage of other foods has been a driver in the domestication of many plant species.

The hop is probably native to much of Europe, including the UK, and its use in brewing is said to be ancient. The first reference to hopped beer is attributed to the Finnish saga the Kalevala, which dates from 3,000 Before Present (BP). Within Britain the use of hops did not become commonplace until fairly recently. Henry VI outlawed the cultivation of hops. Henry VIII forbade brewers to put hops and sulphur into ale, and Parliament petitioned against 'a wicked weed that would spoil the taste of the drink and endanger the people'. Not until the seventeenth century did the cultivation and use of hops in brewing become widespread in Britain.

Hops and cannabis both occur as separate male and female plants. The hop is a tall vine, which climbs clockwise up to six metres through other vegetation in the wild, or in cultivation, up wire frames. Both hops and cannabis plants are covered with resin glands that secrete aromatic oils. These complex chemical properties have been the attraction behind the cultivation of these species. In both crops, years of selection by growers has resulted in cultivated varieties with much higher yields of resin than their wild ancestors. These resin-producing glands are not randomly located about the plants but are concentrated around the flowers, particularly the female flowers. In fact, technically the term Ganja refers only to the flowering tops of the female cannabis plant. Thus, although Jamaican law was amended to extend the definition of Ganja to cover all parts of the plant, it remained technically only illegal to cultivate the female of the species. And since there is no test to determine if the resin is derived from a male or female plant, this provided a legendary loophole.

In hops, the female flowers grow in clusters surrounded by leafy scales, the whole structure being termed a cone or catkin. Hop growers cultivate only the female plants. In Germany, growers have been known to exterminate male plants in the wild to preserve the virginity of their cultivated female catkins. Their reason for doing so is that they consider that the female flowers in seed have a poorer taste than the unfertilized flowers. Furthermore, hop seeds are also reputed to interfere with the fermentation process in bottom fermented larger beers. In Britain, ales are traditionally top fermented with a different strain of yeast, which is unaffected by the presence of a few hop seeds floating in the wort. Thus, British hops, unlike those in the rest of Europe are allowed an uninterrupted sex-life.

The use of hops as a garden herb is known to date from Roman times. Pliny, the great Roman chronicler explains that its Latin name, *Lupulus* is derived from Lupus the wolf, because the plant embraces others as the wolf does a sheep. Since ancient times the plant has been ascribed many different medicinal properties. Its use in hop pillows is now about all that survives of this tradition and is related to its powers to cure insomnia, especially in men. It was also reputed to prevent premature ejaculation, but one cannot help wondering if this is not perhaps a direct result of it inducing sleep. In contrast, in women, hops were said to act as an aphrodisiac. It is difficult to think of a more frustrating combination of powers.

Fig. 3.3 Hops have separate male and female plants. In many areas the males are removed to prevent seed-set as they interfere with the fermentation process.

Avocado – the flasher

Wild plants don't just have weird floral structures like orchids and figs that complicate their love lives and make them problematic to domesticate; they can also do remarkably complex tricks with very simple flowers. A beautiful example of this is the avocado.

Throughout the developed world, the avocado is regarded as a luxury item. As one of very few fruits that are high in fat, being up to 30 per cent oil, it is often considered an extravagance. However, in much of central and southern America the avocado has been a staple of the rural poor for thousands of years. Its everyday usage has resulted in it being termed 'the butter of the poor', vegetable butter and midshipman's butter.

The avocado is a tropical or subtropical tree, which grows to about ten metres tall. It exists as three distinct types, sometimes considered as sub-species. These are: the Mexican, the Guatemalan and the West Indian (which is originally from Colombia in northern South America). These different types may well be the descendants of independent cultivation events from the wild. Archaeological records indicate that humans have been consuming avocados for as long as 9,000 years and have been cultivating them for as much as 7,000. These three types are adapted to the different climatic zones of their origins, with the Mexican being the most cold tolerant and the West Indian being truly tropical. The majority of avocados finding their way to the tables of Europe and North America are of the Mexican type which is smaller and richer in fat than the tropical form. Over time, however, hybridization between these three traditional types is resulting in a blurring of these distinctions.

In addition to being divided into geographic races, avocado trees are split into two types with respect to their time of flowering. The species exhibits a unique flowering behaviour, which is technically referred to as 'protogynous, diurnally synchronous dichogamy'. So what does this botanical jargon mean? All trees have flowers, which look more or less identical, with both female and male parts within them. Nothing unusual here. The odd thing about the avocado is that when a tree flowers, all of its flowers do so at the same time and later that day they all close again. However, in about half the trees, those of the A group, the flowers open first in the morning and only their female parts are mature. Later that afternoon these flowers, which may by then, have already been pollinated by visiting bees, close for the night. The next morning the same flowers reopen. This time their female parts have passed their prime and no longer function, but the male anthers are now mature and shedding pollen.

A second group of trees, the B group, not surprisingly do the reverse. They open their flowers first in the afternoon. At this first opening, these flowers are again functionally female. As night approaches, they too close on mass until the next morning. When the B flowers reopen they have also miraculously changed from being female to become male. The whole elaborate exercise seems to have evolved to increase the chances of cross-pollination, while limiting the amount of self-pollination. Group A flowers are female in the morning and open to pollination from group B which are in their male phase, and by afternoon the situation has reversed. Wonderful, but does it work? Well sort of; certainly isolated avocado trees do not set a lot of fruit. The little there is appears to result from some environmentally induced variation with a few flowers opening out of synchrony with the rest. Even without this cheating the system does not seem very efficient. When groups of A and B trees are grown together less than 0.1 per cent of all flowers actually result in fruit. However, this ratio of flowers to fruit produced is not unusually poor for a tree crop, and presumably the trees are usually successful in avoiding the 'sin' of inbreeding.

Papaya – the sex-change king, queen and more besides

We have already seen that the complex sexual antics that occur within the flowers of orchids and figs are the result of their unusual morphologies. In contrast, the avocado, manages to spice up its sex-life by using very simple flower forms but effectively changing their gender over time. Just imagine how weird your sex-life might be if you were to combine all of these possibilities within one species. This is exactly what has happened in the papaw. The papaw or papaya not only has two well know names, but it is said to have 31 different sexes, and presumably a highly complex social life as a result.

These 31 different sexes can be simplified as males, females and hermaphrodites. Some individual plants always remain predictably male or female or hermaphrodite. However, things are more complicated than that. Some hermaphrodite plants only ever have hermaphrodite flowers, which produce both pollen and eventually fruit from the same flower. Other hermaphrodite plants have separate male and hermaphrodite flowers. While other plants can have separate hermaphrodite and female flowers and some indeed have separate male, female and hermaphrodite flowers simultaneously all on the one plant.

Things start to get really complicated when some male and some hermaphrodite plants go through seasonal sex changes, or respond to physical damage in the same manner. Thus, some male plants will produce hermaphrodite flowers and subsequently fruit at certain times of years, or if their stems are damaged by blows from a cutlass. This technique is often used by those unfortunate enough to find that their solitary papaw plant in the back garden is male and therefore unlikely to produce fruit if left to its own devices. Similarly, a hermaphrodite plant, which normally only produces fully hermaphrodite flowers, may sometimes be induced to produce entirely female flowers as well. Hermaphrodite plants which normally produce only male and hermaphrodite flowers may change to stop producing male flowers and start produce female flowers instead etc. Confused? Just imagine how complex their social life could be!

Then you could ask, how is all this sexual variety regulated? The answer is more complicated still. In fact, there are several competing theories, and it would be true to say we just do not know in detail. However, basically, it appears to be regulated by sex chromosomes similar to those in humans with XX female and XY male. But papaws also have another form of the Y chromosome, such that XY2 individuals are hermaphrodites, and combinations with any two Ys are lethal. With this system, if you cross a male plant with a hermaphrodite you get equal numbers of males, females and hermaphrodites in the offspring. But if you cross a female with a hermaphrodite you only get females and hermaphrodites.

In addition to its incredible love life, papaws have another amazing ability. The latex of the papaw plant contains not one, but two protease enzymes. These chemicals actually digest proteins. Within the plant world this ability is usually only associated with insectivorous plants such as the Venus flytrap and pitcher plants. The papaw must have evolved this ability to discourage grazing animals. After all you are unlikely to consume a plant that is going to digest you rather than vice versa. What is more, we must ask, how did the papaw manage to evolve the ability to synthesize these protease enzymes without digesting itself? This ability to digest protein has independently evolved within pineapples and kiwifruit. The genes involved in all three species are virtually identical. However the protease enzymes differ considerably between species in the degree to which their chemical structures are modified by the addition of sugar molecules.

The latex is tapped from the unripe green fruits by making a series of cuts into the fruit with a piece of broken glass. This operation is performed early in the morning. Throughout the day the latex drips from the fruit into a coconut shell or pot. The sap is then sun dried. About 1,000 fruit are required to produce enough sap to make half a kilo of the dried product. Mankind has found many different uses for the protein digesting sap of the papaw plant. In addition to the obvious use as a meat tenderizer, papaw extract has been used to digest protein hazes out of cloudy beer, employed medically to dissolve unwanted growths, and to remove hair from hides before tanning. It is used widely to remove unwanted protein residues in many manufacturing processes and routinely to purify DNA extracts in modern molecular genetics. Perhaps the grossest use of papaw extract is its injection into cattle before slaughtering. Under these circumstances the protein in the animal's muscles actually starts to break down while it is still alive. However, such meat understandably comes with the health warning – 'do not eat rare!' otherwise you too could be digested by your own dinner.

Cats eyes Cunningham's carrots

While the exotic sex change antics of the papaw are well known to all who try to grow them, some crop plants are a lot more discreet about their love lives. Indeed some of the tricks that are employed by crops to attract pollinators are seductively subtle. The humble carrot is such a beast. The wild carrot which is a member of a family of plants known affectionately by botanists as the 'Umbels' is a master of deception. The Umbelliferae, to give the family its formal title, contains many cultivated species such as celery and parsnip, and the herbs fennel, parsley and aniseed, but also contains many toxic species including hemlock and the blistering giant hogweed. The Umbels are so called because their flowers are grouped into clusters like miniature umbrellas, which in turn are grouped into larger umbrellas to produce a mass of flowers. The wild carrot, which is not uncommon in Britain, can be distinguished from the other Umbels because the central flower of its umbrella is not white like the rest, but is dark purple or pink. This solitary pigmented flower is thought to act as a decoy beetle, tricking passing insects into visiting the mass of flowers and hopefully bringing about pollination. The carrot pulls off this amazing trick by activating an anthocyanin gene that produces a strong red colour only in the central flower. Not surprisingly traditional herbalists spotted this curiosity and regarded these pigmented flowers as possessing special healing properties. Given anthocyanins have antioxidant properties, they may have genuine health benefits, but there are probably more efficient ways of incorporating them in your diet.

The carrot not only has a history of deceiving pollinating insects, it also played a significant war time role in helping deceive the Luftwaffe. A few minutes after midnight on the 19th of November 1940, John Cunningham in his RAF Beaufighter, shot down a Junkers 88 over East Wittering in Sussex. The incident was not only to make the young Flight Lieutenant famous as Cats Eyes Cunningham, it was also to result in those immortal words uttered by so many parents to their children, "Eat up your carrots dear, they will help you see in the dark". Newspaper editors were quick to follow up reports that Cunningham's incredible night vision was linked to his fondness for carrots. The story of how the RAF used such misinformation about night crews being fed large quantities of carrots is now almost as well known as the fact that carrots are good for the eyes. The purpose of this deception was to disguise the real reason behind their improved success in locating enemy planes by using the newly introduced technology of radar. However, as with much effective propaganda the story was not entirely without a basis in fact, and newspaper headlines such as 'Carrots DFC is Night Blitz Hero' were not so far off target.

It had been known since the 1930s that carotene, the yellow-orange pigment that gives carrots their colour is transformed by the lining of the intestines into vitamin A and that deficiency in this compound results in poor night vision. Inspired by this information, war-time agriculturists managed to develop new high-carotene carrots containing two or three times the amount of carotene found in conventional carrots. The amount of work involved in this task makes it likely that this was genuinely intended to help minimize night blindness in aviators, rather than being part of the elaborate misinformation surrounding the introduction of radar.

Whatever the truth behind the tales of carrot crunching night pilots, what is certain is that this was not the only role that this humble vegetable was to play in the British war effort. With sugar in short supply, slices of carrot having high sugar content were incorporated into sweet pies and flans. The roots were boiled down and reduced to make carrot jam, or roasted until black to make a coffee substitute. However, the carrot has not always been so sweet, or even orange. The first carrots to be cultivated appear to have been introduced to Europe from Afghanistan. These eastern carrots were dark red to almost black in colour, being pigmented by compounds called anthocyanins which also give red wine its distinctive colour. Like beetroot, purple carrots exude their colour during cooking, turning stews and soups a nasty brownish purple. By the mid nineteenth century, modern anthocyanin free yellow and orange carrots were developed in Holland. These rapidly replaced the older red types, which are now virtually extinct.

Einstein and the bees

The mysterious decline in honeybee populations across much of the world has been splashed across the media as a portent of doom. Frightening statistics are bandied around. The UN claim that, "Seventy out of the top 100 human food crops, which supply about 90 per cent of the world's nutrition, are pollinated by bees" and also that this is worth 134 billion pounds per year. Albert Einstein is reported to have said that, "If the bee disappeared off the surface of the globe, then man would only have four years of life left. No more bees, no more pollination, no more plants, no more animals, no more man." It is perhaps not a surprise to discover that this quote is apocryphal, because Einstein is famous for the equation $E = mc^2$ rather than Bee $= mc^2$!

Although the decline in bee populations is alarming, many of the statistics reported in the news have been carefully selected to over dramatize the situation. I have argued in this chapter, that while many wild plants have complex pollination mechanisms, we have for the most part tended to avoid domesticating such species because this habit may limit their ability to produce fruit in the absence of their co-evolved pollinator. Thus, most crops can be pollinated by many generalist species of insects, and this has allowed them to be successfully cultivated around the world and be visited by whatever insects are available locally. In addition to this, most of our staple crops; wheat, maize, and rice are grasses that rely on the wind for pollination, while others such as potatoes and yams are spread vegetatively and are rarely propagated from seed at all. Even crops that would have originally been insect pollinated, such as oil seed rape (which has bright yellow flowers and produces lots of nectar to attract insects) become wind pollinated when they are grown on an industrial scale. Ask anyone who suffers from hay-fever!

Although the majority of our staple crops are wind pollinated, it is striking that most temperate fruit trees are insect pollinated. This is particularly strange since the majority of these trees originally evolved in deciduous woodlands that are dominated by large wind pollinated species such as oak, ash, beech etc. In their natural habitats, the ancestors of apples, pears, cherries, almonds, peaches etc., are never as abundant or as tall as the wind-pollinated species. There is a simple biological explanation responsible for this fact. Wind pollination is an effective mechanism to ensure fertilization if you are abundant. But smaller, less common species cannot rely on such a random delivery method and have to utilize the more precise pollen delivery service operated by insects.

This observation begs the question; why are so few of our crops derived from wind-pollinated trees? Acorns could make a perfectly acceptable nut, but yet we have domesticated insect pollinated almonds instead. This comparison has been made before and the following explanation proposed: As long-lived plants both oaks and almonds are full of chemical defences to prevent herbivore damage. However, the cyanides found in almonds are controlled by a single gene and thus it is easier to select for non-poisonous almonds than it is for edible acorns, because many genes regulate toxicity in oaks. However, this theory seems unlikely as acorns are easily detoxified by boiling and have historically been eaten as both a meat and coffee substitute by many humans. There is in fact a very simple but overlooked reason why non-toxic acorns have not been developed. If our ancestors had been lucky enough to find an edible non-poisonous oak tree, its flowers would have been swamped by pollen from the abundant wild toxic oak trees that surrounded them. This explains why although there are wind-pollinated cultivated hazels, the nuts they produce are remarkably similar to the wild cob-nuts found in our hedgerows. In contrast, once you have identified an almond that is not packed with cyanide, it is easy to establish an orchard of these trees because the pollinating insects are likely to ensure that they only pollinate each other rather than cross with the few rare wild toxic almond trees that are found deep in the forest. This also explains why tree crops are much more common in tropical than in temperate agricultural systems. Tropical forests are famously high in diversity. Here no single tree species dominates. Typically each hectare of tropical forest will only contain two or three individuals of each species of tree. Wind pollination is simply not an option. Under these conditions genetic isolation from wild individuals can easily be achieved. The early generations of domestication can therefore diverge from their wild ancestors much more rapidly than is possible for abundant wind pollinated trees. Thus, there are many tree crops that are pollinated by generalist insects such as bees, but these tend not to be the staple crops that provide us with the nutrition that we need to survive. These are typically the fruit crops that make life more pleasurable, such as apples and cherries and are important in supplying the vitamins that keep us healthy. The plants that fill our larders with the stores we need to survive are generally ones that evolved food storage organs of their own. Domesticating these plants presents us with another set of problems that will be considered in the next chapter.

References

Most significant new sources in the order they were utilized in this chapter.

Richards, A.J. (1997) *Plant breeding systems*. Second Edition. New York: Garland Science.

Cameron, K. (2011) *Vanilla orchids: natural history and cultivation*. Oregon, USA: Timber Press, Inc.

Irwin, R.E. and Brody, A.K. (1998) Nectar robbing in Ipomopsis aggregata: effects on pollinator behaviour and plant fitness. *Oecologia*, 116: 519-527.

Kjellberg, F., Gouyon, P.H., Ibrahim, M., Raymond, M. and Valdeyron, G. (1987) The stability of the symbiosis between dioecious figs and their pollinators: a study of Ficus carica L. and Blastophaga psenes L. *Evolution*, 41: 693-704.

Stewart, A. (2013) *The drunken botanist - the plants that create the world's great drinks*. North Carolina, USA: Algonquin Books, Workman Publishing.

Purseglove, J.W. (1968) *Tropical crops: dicotyledons*, Volume 1 & 2. New York: Wiley.

Heywood, V.H. (1983) Relationships and evolution in the Daucus carota complex. *Israel Journal of Botany*, 32: 51-65.

4

Storing up trouble

Why do so few crops provide us with most of our calories? It is remarkable that we eat very few species of the plants that are available. However, it is even more astonishing that we gain most of the energy we need to survive from a tiny sub-set of these species. This chapter explores this conundrum and finds that energy rich plants tend to contain toxins, and perhaps surprisingly, therefore, some of our most important crops have a propensity towards being poisonous.

In our modern world it is easy to be blissfully unaware of the most important challenge that species face. Although the solution to this problem may now seem trivial, for most of our history we have shared this challenge with plants and animals alike. In solving this dilemma for themselves, plants have frequently also provided us with a solution, while simultaneously creating a whole new set of difficulties for us to deal with. The conundrum is ensuring that you have enough food to survive through lean seasons. The evolutionary struggle to eat and avoid being eaten has been highly influential in determining which plants we have domesticated, with different groups of crops providing us with sustenance and others helping protect this food from other hungry species competing to consume the same stores.

Around most of the earth its inhabitants experience environmental seasonal variation. Confronted with harsh conditions there are a number of things you can do. Firstly you can migrate to somewhere more hospitable. Secondly you can reproduce and die; in the hope your offspring will stay dormant until the good times return. Or finally, you can hibernate and lay dormant. For this last option to be effective, it matters not if you are plant or animal, the important thing is to have sufficient stored resources to keep you alive. Employing either strategy two or three is likely to make any wild plant an attractive target for domestication. An annual lifecycle involving the mass production of seeds followed by the death of the mother plant is found in many of our crops. This life-style choice is however, not that common in the wild. Here, long-lived perennials tend to out-compete annual species in all but the most disturbed of sites. Thus, wild annual plants tend to be associated with accreting sand dunes, new volcanic soils, retreating glaciers and other ephemeral environments. Here there is open ground where seedlings can grow rapidly without encountering older bigger plants. Humans are good at creating such disturbed areas, and thus annual weeds follow us around and are frequently associated with human habitation and agricultural activity. But in our absence, relative stability reigns and then perenniality is the order of the day.

In many undisturbed habitats there can be periods of time when conditions are not ideal for plant growth. In temperate conditions we tend to associate this with the winter. In contrast, in Mediterranean climates it may be too hot and dry in the summer and more suited to plant growth in the cooler wetter winter months. Here plants will often spend the summer months below ground hiding from the sun. Similar pressures are associated with the ground flora of deciduous woodlands. In the summer, beneath the shade of a broadleaved canopy, very little light reaches the forest floor. During these months, the low growing woodland plants simply cannot photosynthesize. These species are forced to exploit the brief period of early spring to grow and flower, before the trees are in full leaf. This is why bluebells, violets, wood anemones, primroses etc., all bloom in early spring. Even in tropical regions where seasonality can be less marked, there can be dry seasons where the vegetation withers and apparently dies.

Whenever perennial plants face seasonal hard times, their survival strategy frequently involves a period of dormancy, with many retreating back to a vegetative structure and laying quiescent until conditions improve. These food storage organs may be formed by adapting roots in the form of tubers, or swollen as taproots. Stems may also be modified as tubers, rhizomes or corms. In lily-like plants, even leaf bases may become swollen and act as the storage organs that we recognize as bulbs. All of these food crammed structures must have attracted the attentions of our hungry hunter-gatherer ancestors and also encouraged early human farmers to propagate any leftovers, in the certainty that bad times were just around the corner. It is no surprise therefore that many different plants with such adaptations have been domesticated on many occasions in human history and are important in our diets to this day.

If only life was so easy. Plants that store food reserves below ground in tubers, rhizomes and bulbs etc., would not just attract the attention of hungry humans, a host of other local creatures would be driven to try to exploit these goodies too. As we shall see, the standard evolutionary response in these plants to the threat of being eaten has been to deploy a range of highly toxic defences. Thus, we shall discover that a recurring theme in the domestication of these important crops has been how to deal with these poisons.

Not only do plants that cower away below ground in unfavourable seasons face the unwanted attention of animals determined to eat them, they are also forced to have a sex-life that is short in duration. An active but brief flowering period does not come without risk. For example, seed-set in spring flowering woodland species can be erratic. In some years all will be well. However, these marginal seasons can be unpredictable. Beautiful spring days can be interspersed with frosty nights that kill not only tender flowers, but also the pollinating insects that they depend upon. In such years, seed production may fail. However, evolution has again provided these tenacious plants with a rather cunning plan B. If these first flush of blooms fail to produce viable seeds, many spring flowering, woodland species have the ability to produce a second set of flowers. Unlike the first familiar set of flowers, these secondary blooms often go unnoticed. They typically appear as flower buds, and never open. Thus, the delicate sexual organs of the plant are kept snug and warm, away from the frosts and winds of early spring. These cleistogamous flowers (to use their technical name) automatically self-pollinate and thus don't require the services of unreliable early spring insects. Of course, the fail-safe seeds that these flowers produce will be inbred, but better self-pollinated seeds, than fail to produce any offspring, just because spring was cold and wet.

The co-evolved adaptations of producing tempting food storage organs and cramming flowering into a short season could be problematic in terms of domestication. As we saw in the last chapter, such unusual pollination mechanisms may fail to function effectively if they are transplanted away from familiar environmental conditions and insects. But of course, the sexual practices of this group of plants are irrelevant. Farmers simply don't care if their potatoes, yams etc., flower or not, because they are nearly always propagated asexually by planting spare tubers.

The bitter, sweet, life giving and poisonous cassava

Cassava is one of the world's most important staple crops and yet it is highly poisonous. So when we are considering why we eat so few plant species of the 300,000 possible edible ones, it is even more remarkable that we have decided to consume plants that might even kill us. Cassava does not have a good pedigree; it is a member of the spurge family, which contains 5,000 species, many of which are highly toxic and includes no other food crops. The mancineel tree, which grows along the beaches of the Caribbean surrounded by notices warning tourists not to touch, is a close relative and one of the most poisonous trees on earth. Its Spanish name translates as "little apple of death". Another relative, *Croton tiglium* provided pharmacists with the most potent of all known purgatives. However, its use has now been prohibited because taking it made unfortunate patients feel like the bottom was falling out of their world (or vice versa)! More benign, but still toxic members of the spurge family include; the rubber tree, poinsettias that are so popular at Christmas and jumping beans, whose seeds entertain young children by moving erratically when warmed. The jumping ability of these beans is a result of them containing moth caterpillars that might be suffering symptoms akin to those who have consumed *Croton tiglium*. Cassava itself contains cyanogenic glucosides. When its tissues are damaged these react to produce hydrogen cyanide. If consumed raw or improperly prepared, the symptoms of cassava poisoning can include: vertigo, vomiting, partial paralysis and potential death within a few hours.

With this background why would anyone think of cultivating cassava? The answer is of course that its root tubers are a highly valuable source of carbohydrates. The plant, which is a shrub that can grow to three metres tall, is resistant to drought conditions and being highly toxic suffers from few serious pests. In addition to this it is easily cultivated as it can be readily propagated by planting 15 centimetre stem cuttings. It was probably first domesticated 10,000 years ago in Brazil. In this region its wild relatives can still be found. But cassava itself is only known in cultivation or as a recent escape. The first definite archaeological evidence of cassava consumption comes from Mayan sites in Mesoamerica which are much more recent. By the time of the Spanish colonization of the new world, the crop was being wildly grown and eaten throughout much of Southern and Central America and the Caribbean.

After thousands of years of domestication, varieties of cassava have been developed that contain lower levels of cyanide. These sweet varieties are now widely grown throughout the tropics where some half a billion people rely on them as their staple food. In addition to producing less toxic varieties, humans have developed a number of different methods for extracting the cyanide from cassava tubers. These methods involve, drying and grinding into flour, soaking in water, cooking and fermenting. It can take more than 24 hours of soaking in water to reduce cyanide to safe levels. Even then there is a degree of risk involved, because even sweet varieties grown in drought conditions can contain elevated levels of poison. Once processed, there are a variety of ways that cassava can be eaten. Cassava flour can be baked into bread and cakes, boiled as more substantial dumpling like balls, or made into thick porridge like soups. The roots may be sliced and fried like potato crisps. Outside the tropics, perhaps the most familiar but unloved of all cassava recipes is tapioca pudding. Boiled with milk, tapioca pearls form a gelatinous mass that has terrified many a young school child at dinnertime. Why anyone would want to subject children to this delight is more of a mystery than why farmers still widely cultivate bitter varieties of cassava. These more primitive varieties can contain 50 times more cyanide than the sweet cultivars. Yet farmers still grow them, because they tend to be more drought resistant, suffer from fewer pests, and not surprisingly are less likely to be stolen. This phenomenon has been repeated with many crops. Through the domestication process, we have increased their palatability by selecting out the toxic chemicals that evolved to protect them from herbivores. In making them more appetizing for ourselves we have rendered them more susceptible to pests and diseases. The final ironic piece of the jigsaw is that we are now frequently forced to develop our own chemical defences. We then bombard our crops with these pesticides to replace the chemicals we have spent generations selecting out. The question that divides many agriculturalists is which is most damaging to human health, the naturally evolved toxic chemicals that our crops would have originally contained, or the chemical pesticide residues that have superseded them?

The nature of crops

Fig. 4.1 Cassava is one of the world's most important staple crops in spite of containing high levels of cyanide.

I want yams where I am

The many different and unrelated crops that have been domesticated for their starch filled storage organs have evolved different chemical defences to protect themselves. For example, wild and green potatoes contain poisonous alkaloids, taro roots are stuffed with calcium oxalate and as we have seen, cassava is full of cyanide. Such plant chemical defences have generally been a problem to those attempting to either domesticate or exploit species containing them. On other occasions we have discovered novel ways of utilizing these compounds for our own ends. For example the tubers of yams have been harvested for their toxins to poison arrows, to poison in-laws, to catch fish, and to apply to other crops as an insecticide. In more recent times, large quantities of wild yam tubers have been collected commercially for an even more unusual purpose.

Yams grow wild all around the world, scrambling through other vegetation, with some vines twisting clockwise, and others anticlockwise, depending on the species. Plants are usually single sexed, producing either all male, or all female flowers, but with both types producing large subterranean tubers. Wherever yams have occurred, they appear to have attracted the attention of man, with different species being domesticated in Asia, Africa and South America. The greater yam, the lesser yam, the Chinese yam and the bitter yam were all first cultivated in Asia. The yellow and white guinea yams and cluster yams are of African origin. The potato yam, which produces strange aerial tubers, occurs wild in both Asia and Africa and appears to have been cultivated independently in both regions. In South America only the cush-cush yam was considered worthy of domestication, but other species may have been collected from the wild. Even Britain has a native yam, called black bryony. Although it is now generally considered to be toxic, bryony tubers, measuring up to 60 centimetres, were eaten from prehistoric times. Even to this day some French cookery books describe how to detoxify them by prolonged soaking and boiling. The French name for the plant is herbe aux femmes battues, or literally the battered wives' herb. Within the UK and the US the tuber most commonly called a yam is in fact the unrelated sweet potato, which is actually related to the ornamental morning glory.

The majority of the world's yams are grown and consumed in the 'yam belt' of West and Central Africa, in the region east from Ivory Coast through Ghana, Togo, Benin and Nigeria to Cameroon. In spite of yam cultivation being labour intensive and yields being low, yams have been particularly valuable in the tropics because they are easy to handle and store well for several months. Their keeping properties made them ideally suited for ships' stores in the days of sail. In this manner yams were transported from Africa to the Americas as provisions on board slave ships and similarly Asian yams were dispersed across the Indian and Pacific Oceans in the pre-European period.

Within the last one hundred years wild yams have been harvested for a new and unexpected purpose. The story dates from 1933 when the Eastman Kodak Company first isolated the human hormone progesterone. Huge quantities of cattle brains were required to produce a few grams of the desired compound. It was quickly realized that progesterone had great potential as a contraceptive as it was found to halt ovulation in laboratory rabbits. The discovery that the hormone could be synthesized from various naturally occurring plant compounds stimulated a search, which was described by an American pharmacology newsletter as a 'story warranting a Hollywood movie, full of intrigue, deception, scandal, corporate envy, trickery, bribery and even violent harassment and murder'. Although 250 plants containing progesterone like compounds were identified, including: red clover, liquorice, fennel and soybeans, the first commercial synthesis of progesterone for use as an oral contraceptive was from diosgenin, derived from wild Mexican yams. This simple fact appears to have generated a plethora of stories and claims about the medicinal properties of wild yams.

Lotions and potions containing wild yam extracts or 'natural progesterone' are marketed as alternative medicines as a form of hormone replacement therapy, or to ease the discomforts of pre-menstrual tension, and even as contraceptives. Stories are told of South American Indians, who for centuries were able to regulate their own fertility by chewing on yams. The truth behind all these tales is that the compounds found within yam tubers are grazing deterrents and not active human hormones. Although yams have been used in the synthesis of the Pill, there are several complex chemical changes involved in the conversion of the naturally occurring diosgenin to progesterone. These are chemical changes that the human body is unable to perform for itself. If you were considering eating yams as a form of birth control, you may just as well have a box of liquorice allsorts or rely on a lucky four-leafed clover.

Learning from the potato

Historians tell us that the reason for studying the past is to prevent us from repeating the mistakes of our ancestors. The rest of us know that its true purpose is to allow us to smile smugly at the follies of forebears. Proof of this, if proof were needed, is the potato, which more than any other crop has attracted the attentions of historians. So what have we learnt from the potato induced disasters of the past?

Potatoes were first domesticated about 7000 years ago in the Peruvian High Andes around Lake Titicaca. In this region wild species of potato can still be found. They include highly variable sexually promiscuous species and less variable inbreeding species. Occasionally tubers are still harvested from the wild, but as with so many of our storage organ based crops the wild types are extremely bitter and rendered virtually inedible by toxic compounds called alkaloids. These same chemicals are also responsible for the poisonous nature of a relative of the potato, the deadly nightshade. The first step in the domestication process must therefore have been the selection of less bitter, less toxic varieties. This was achieved with some success, but it is frequently said that if the potato were to be invented today, it would be banned because of the toxic residues it contains. The next stage of domestication involved a doubling of the number of chromosomes. Potato scientists cannot agree if this occurred automatically or following the hybridization of two different species. But this matters little to the story.

Many historians and comedians have spun wondrous and apocryphal yarns about the introduction of the potato into Europe. Sir Frances Drake is said to have discovered them in the Caribbean whilst returning from evacuating a failed British settlement in Virginia. Considering them too good for Queen Elizabeth, on his return Drake passed his potatoes on to Sir Walter Raleigh who had them planted on his estate in southern Ireland. By 1590 they were ready to harvest. Unfortunately Sir Walter tried to eat the toxic potato fruit. Disgusted by the experience, Sir Walter ordered the plants destroyed, and only then when his gardener took a spade to the plants were the potato tubers finally found. Of the details included in this version of the story, only the date is anything like accurate!

The Spanish were the first to introduce the potato into Europe in around 1570. Twenty or so years later they were independently introduced into Britain. Both introductions were from the South American Andes. In spite of the fact that the British frequently referred to the crop as the potato of Virginia, it was not known in North America or the West Indies until being introduced via Europe in 1621. Another inaccuracy in the tale of Drake and Raleigh is the claim that their potatoes were able to produce tubers. The first plants brought into Europe, being from the Andes, were not adapted to the long summer days of the temperate North. Andean potatoes are stimulated to form tubers by the short days of more tropical climes. It was to take almost 200 years of selection before the new arrival became adapted to European long summer days. The fact that plants with storage organs have evolved to cope with specific seasonal climatic variation means that they can often be problematic to successfully cultivate in different regions of the world. This fact might help explain why to this day there are so many similar, minor crops that are only cultivated within rather limited geographic regions. This list could include: jicama, oca, water chestnuts, taro, arrowroot, rampions and the lotus. Even with some more widespread crops such as the onion, plant breeders still struggle to select varieties suitable for different climate conditions and longer seasons.

The time taken for the potato to acclimatize to northern European seasons limited its uptake by the locals. In addition, and as we have already seen with its close relative the tomato, there was a reluctance to eat potatoes because they closely resemble the poisonous nightshades. These factors combined to prevent the widespread growing of the crop in Britain until about 1800.

Partly because it thrives in mild wet conditions and also because their English landlords kept much of the best land capable of growing cereals for themselves, the Irish took to potato cultivation in a big way in the early eighteenth century. The crop was grown in long strips of ground about two metres wide, along which manure, seaweed or rotted turf had been heaped. On top of this, earth taken from between the strips, was piled. The potatoes were planted with a dibber or placed on top of the manure layer during construction. Either way ensured that they were above ground level and so protected from water logging. The method was so effective that with a few other basic supplies, a strip of land just 700 metres was all that was required to support a family. The beauty of the system was that because it kept the potatoes frost-free it also worked as a store. Once planted these plots required no further attention until the day came to lift them for consumption. For this reason, the Irish strip method of growing potatoes became known as 'lazy beds'. The success of the crop enabled the population of Ireland to increase dramatically, with estimates claiming a 900 per cent rise from one million to nine million in the hundred years from 1740.

However, history records in a series of grim statistics that potato cultivation in Ireland was to end in disaster and in the process change the world. Although the Irish potato famine is associated with the years 1845 and 1846, the previous one hundred years had seen a procession of nearly thirty different famines. Each one was caused by the destruction of the potato harvest due to a fungal, bacterial or viral disease. Probably as many as half a million or one third of the population died around 1740. During the blight years of 1845 and 1846 one million people died and a further million and a half emigrated. This set a trend, which resulted in more than five million leaving the country before the first decade of the twentieth century. Potato cultivation played a similar role in the Highland clearances of Scotland and the mass emigrations or ethnic cleansing that followed.

With hindsight, historians and agriculturists have both retrospectively been able to predict the catastrophe. Such heavy reliance upon a single crop with only low levels of genetic diversity in a mild damp climate is a recipe for disaster. The structure of the potato's lush dense foliage seems to invite attack from diseases. As an agricultural system it could have been designed to propagate pathogens.

So have we learnt the lessons of the potato? Global production of the crop is steadily increasing, and is still based on a very narrow genetic base. It is the only non-cereal in the top eight crops, which dominate world food supply. This reliance on so few crops is a particularly risky strategy given the uncertainties of climate change. Predictions vary, but generally forecasts include increases in both temperature and precipitation i.e. the climate looks like becoming more favourable for plant diseases. It would strike you as perhaps a good time to be storing away food reserves just in case one of the eight major staple food crops was to suffer an attack from a new 'blight'. Just the opposite has occurred; set-aside policies and the pressures of free trade have seen the depletion of the food mountains of the 1980's. At the time these were seen as the obscene symptoms of a crazy agricultural policy. However, with food security becoming a real concern in the twenty-first century, perhaps we should be thinking again of storing food for uncertain seasons ahead, in just the same way that the potato has done for millennia.

On the positive side plant breeders now better appreciate the value of genetic resources, and great efforts are made to collect and conserve the genes of all the world's major crops. For both historical and political reasons the Centro Internacional de la Papa, the international potato centre and world potato gene-bank is located in Peru. For years the research centre has been regarded as a legitimate target by terrorists. So next time you eat a plate of chips, spare a little thought to the group of international potato scientists, who live their lives behind two sets of barbed wire and armed security guards to ensure that it is not chips for the future of the potato.

Hormone replacement therapy and the sexing up of taro

We have already seen how the stunningly beautiful flowers of the orchids and the weirdly bizarre blooms of the figs have probably contributed to restricting their utility as crop plants. However, the largest, smelliest, and arguably the strangest flowers of all belong to a family of plants that have provided us with a number of food crops. These are the aroids. It is not just their weird flowers that make the arums unlikely candidates for domestication. As with many other crops that have starchy storage organs, most of the aroids are defended by toxic chemicals. In this case the poison of choice is calcium oxalate. Although this substance can be fatal to humans it is found in many of our familiar crops, such as tea, kiwifruit, spinach and rhubarb. It also turns up as the scaly deposit found in the bottom of beer barrels. The good news is that calcium oxalate readily breaks down on cooking, so the starchy corms are easily rendered safe to eat and thus aroids are transformed into potentially ideal candidates for domestication. This probably explains why there appears to have been at least four separate sets of aroids developed as very similar crops. This has occurred independently in Asia, Africa, South America and Oceania, which has given rise to taro, dasheen, eddoes, cocoyam, tanier, swamp taro and giant taro, with many of these names being interchangeable in different regions. Not only are the starch filled corms of these plants widely used as a staple, their fleshy leaf stalks and distinctive arrow shaped leaves are also eagerly consumed as green vegetables in their own right. However, their distinctive arrow shaped leaves should not to be mistaken for arrowroot, which also provides an excellent supply of fine-grained starch, but this is a relative of ginger.

All of the aroids have flowers that could be described as being a bit eccentric. The flower of the titan lily or titan arum is the largest of all known flowers and can reach three metres tall. Its Latin name *Amorphophallus titanum* literally translates as misshaped phallus, which describes a feature common to aroid flowers. In the heart of the aroid inflorescence is a large central spadix. This is a distinctive feature of members of the family. This organ comprises of small, often sterile flowers located around a phallic like structure. In bands below these sterile flowers are often zones where the flowers are male, and below this a band where the flowers are female. These amazing inflorescences tend to be short lived. During this brief period several species have a spectacular trick up their sleeves. In *Amorphophallus* and some other aroid species, including the cuckoopint that is common across much of Europe, the plants are able to significantly elevate the temperature of their spadix until they become distinctly hot to the touch. This feat is achieved by hijacking the metabolic pathway that generates energy within living organisms, so that it produces heat rather than chemical energy. This process may be more demanding of energy than are the beating wing muscles of a hovering humming bird. It is no great surprise therefore, that these flowers only remain hot for a brief period of time. During this time they release volatile, often very pungent chemicals that attract pollinating flies. This is why *Amorphophallus* is also known as the corpse flower. Some of the more elaborate aroid inflorescences not only attract insect pollinators by this unsavoury, smelly method, they also entrap pollinating insects in a structure that resembles a pitcher plant. Newly arrived insects pollinate the receptive female flowers. Before they are granted parole, the insects are liberally showered in pollen from the male flowers that mature during their short spell in custody. Such complex pollination mechanisms may be fascinating, but could be potentially problematic when it comes to domestication. Can you imagine the stench of a field of corpse flowers? But of course we don't require our cultivated aroid crops to flower at all, since we are only interested in harvesting their starch filled corms. In fact, plants that don't bother to flower at all and instead put all their energy into the vegetative growth may make ideal crops. Here lies a Catch 22 problem when it comes to identifying potential plants for domestication.

Many crops including aroids are grown for their vegetative parts. These may be roots, stems or leaves. In each case we don't require that the plants produce a flower. In fact, in many crops, blooming can be most undesirable, as this process may convert the plant from a vegetative growth habit into a flowering one. When this happens the vegetative parts may no longer be produced or may become inedible. For this reason farmers have been selecting over many generations crops that are reluctant to flower or don't flower at all. In several aroid crops they have achieved success. Under most conditions a number of these crops remain totally vegetative and are entirely propagated by planting corms. That is all well and good, until you need some seed. For example, if you want to cross two varieties, to introduce disease or drought resistance. This is a not an uncommon dilemma for plant breeders. How is it possible to breed a non-flowering crop? In taro the answer is a relatively simple one. Non-flowering plants can be coaxed into flowering and thus an active sex-life by the application of a plant hormone called gibberellic acid. This has to be applied in such high doses that there is a danger that it can burn the leaves. In taro plants which are just a bit shy and reluctant to flower regularly or reliably, these can be enticed by much lower hormone concentrations.

Fig. 4.2 Taro is an aroid; the poison of choice in this family of plants is calcium oxalate.

Given how difficult it can be to cajole aroids into having sex and how weird their flowers are, it's no great surprise that we tend not to cultivate them for their fruit. There is however, one unexpected exception to this rule; the Swiss-cheese plant. This stalwart of the 1970s foyer, if given the opportunity to escape the office and procreate, produces a fruit that tastes rather like a mixture of banana and pineapple. Originally native to the humid forests of southern Mexico and Guatemala the Swiss-cheese plant is now grown for its fruits throughout the wet tropics. In this more natural setting it escapes its cheese related label and is known by a range of even more unusual names such as the ceriman, piña anona (Pineapple custard) and arpón común (the common harpoon).

Akees – a national fruit that kills its citizens

Plants don't just need to defend their starchy storage organs from munching herbivores. Seeds also tend to be loaded with precious food reserves to drive germination and fuel early seedling growth. Such seed food stores can prove irresistible for hungry humans and animals alike. It is no great surprise therefore that many seeds are also often defended by a range of toxic chemicals. As with other poisonous storage organs, this can be a less than ideal characteristic in a food plant. In most crops, many generations of selection have reduced such toxins to safe levels. However, consuming a few crops can still be like playing Russian roulette as their fruits may contain fatal levels of poison.

Although it is native to the west coast of Africa, the akee is the key ingredient in the national dish of Jamaica - salt-fish and akees. The akee tree was first introduced to the island by Thomas Clarke aboard a slave ship in 1778. However it carries the Latin name *Blighia sapida* in honour of Captain Bligh (of Mutiny on the Bounty fame) who in 1793 transported a specimen to the botanic gardens at Kew where it was first formally described. Curiously, the other main ingredient of this rather delicious dish, the salt cod, is also imported, typically from Canada. Within a few years of its introduction there were reports of something called Jamaican vomiting sickness. Although until 1875, this was generally attributed to malnutrition, yellow fever or other endemic diseases. For many years there were incidents of the sudden extreme sickness of children and even deaths of entire families. Eventually the problem was traced to two water soluble amino-acids called hypoglycin A and B. These compounds interfere with the human body's energy metabolism, ultimately reducing the amount of glucose in the blood, hence symptoms that resemble malnutrition. These poisons are found in the unripe fruit and also in its glossy black seeds. Like many fruits, akees are adapted to be eaten as a dispersal mechanism. However, for a plant it is not a great idea for its seeds to be eaten before they are mature and so many unripe fruits are cryptic green and poisonous. During the ripening process fruits become more appetizing by increasing their sugar concentrations. Softening they signal this change to nearby animals by changing colour, often from green to more visible red and smelling 'ripe'. In many plants, ripening also involves detoxification of poisons that helped protect the unripe fruit until their seeds have matured. In the akee the process is a simple one. As the fruit matures it bursts open revealing three bright yellow arils and associated seeds. Once exposed to the air, the toxins evaporate. Unfortunately many Jamaicans are just a little too impatient for this process to have totally dissipated the hypoglycins. In their eagerness to consume their national dish, unripe akees are often picked, with fatal consequences. Once ripe, the fruit looks and tastes rather like scrambled eggs and you might think it a strange thing to risk death for. But many a proud Jamaican might beg to differ.

Fig. 4.3 Akees are part of the national dish of Jamaica (salt-fish and akees) in spite of the fact it regularly poisons its devotees.

Onions – a crop worth crying over

In spite of the fact that bulbs are a relatively common storage organ in monocotyledon plants, relatively few of these species have become important crops. The onion and its close allies, garlic and leeks are obvious exceptions. This fact is even more remarkable considering the following contradictory observations. Onions and garlic are amongst the most cosmopolitan of all crops and seem to have been integrated into the local cuisine in every country on earth. At the same time, onions are famous for making cooks cry and garlic for making its aficionados smell! As with other storage organ crops, the potent chemicals that cause these properties must have developed as grazing deterrents. In such cases, domestication typically seems to have removed or reduced the efficacy of the plants chemical defence shield. With onions, the fact that they make us cry has not reduced their popularity in the kitchen. But then crying is rather less significant than actually being poisoned.

The common onion was probably one of the earliest of crops to have been domesticated, because it is easy to grow and stores well - but not forever. Because of their fleshy nature, onions don't leave much in the way of direct archaeological remains. So unfortunately, there is little in the way of hard evidence available for those interested in onion ancient history. Onions are thought to have been cultivated for more than 5,000 years. Although they have left few plant remains, onions have left a strong cultural record, perhaps because of their pungent nature. There are references to onions in India's oldest known texts and their cultivation is described in Samarian documents dating from 2500 BC. The ancient Greeks reported that Olympic athletes were fortified by drinking onion juice and by rubbing onions into their bodies, which must have made drug testing a whole lot easier! The Roman authors, Hippocrates, Theophrastus and Pliny all describe an impressive range of different forms of onions. The most impressive direct evidence we have of Roman cultivation comes from the doomed gardens of Pompeii. Within the lava flows empty rows of distinctive conical casts were found left by the onions that were burned alive along with the city's human residents.

Of all the ancient civilizations, the Egyptians were the most enamoured with the onion. Indeed, onions were worshipped since their concentric rings were thought to signify eternal life. This belief led to the common practice of adorning dead bodies with squashed onions during the embalming process. The mummy of pharaoh Ramses IV appears to have been entombed with onions set into his eye sockets. Egyptian priests were frequently pictured holding bouquets of onions and with altars similarly swathed. The walls of tombs in both Old and New Kingdoms were decorated with pictures of onions as funeral offerings or as part of great feasts. It is perhaps no surprise therefore that the bible records that on escaping from slavery in Egypt that the Israelites lamented the loss of the onion as they wandered in the wilderness (Numbers 11:5).

Five thousand years of such well documented cultivation begs the obvious question, why are we still so enthusiastic about a plant that burns our eyes to the point of tears? The answer seems to be, that our love affair with the onion is in spite of its pungent properties rather than because of them. Recent genetic research has identified that there is an entirely different biosynthetic pathway that produces the sulphurous compounds that make us cry from the process that gives onions their distinctive and highly desirable flavours. It is now therefore possible to develop tear free onions without impairing their flavour. This is another onion oddity. In contrast, the next chapter explores plants, which appear to have attracted our interest specifically because they contain compounds that burn, sting, or have powerful flavours.

References

Most significant new sources in the order they were utilized in this chapter.

Hillocks, R.J., Thresh, J.M. and Bellotti, A. (2002) *Cassava: biology, production and utilization*. Wallingford, UK: CABI publishing.
Romans, A. (2013) *The potato book*. London: Frances Lincoln Limited.
Purseglove, J.W. (1972) *Tropical crops: monodicotyledons*. London: Wiley.
Adams, C.D. (1971) *The Blue Mahoe and other bush: an introduction to the plant life in Jamaica*. Singapore, New York: McGraw-Hill Far Eastern Publishers.
Eady, C.C., Kamoi, T., Kato, M., Porter, N.G., Davis, S., Shaw, M. and Imai, S. (2008) Silencing onion lachrymatory factor synthase causes a significant change in the sulfur secondary metabolite profile. *Plant Physiology*, 147: 2096-2106.

5

The weird and wonderful

Although we gain most of our calories from a remarkably short list of species, our spice-racks, and medicinal and recreational drug cabinets in contrast are stuffed with the plants that are rich in an amazing diversity of chemicals. This chapter describes a range of the more unusual plants that we consume and tries to answer: why are we repeatedly drawn to minor crops that burn our lips and befuddle our brains?

It can be argued that all plants with the exception of grasses can be considered to be poisonous. The cells of plants contain a vast array of weird and wonderful chemicals. The biological function of many of these within these plants remains uncertain. However, once inside the human body these compounds may act as powerful drugs, potentially enhancing our health by mopping up cancer causing reagents, or they may help kill disease-causing microorganisms. Alternatively nature's apothecary may impair our well-being. Plants may cause our demise by rapid poisoning, or slowly cause our death over decades. Other chemicals derived from plants cause allergies, intense burning sensations and befuddle our minds. Therefore, as early humans experimented with which plants to cultivate and which to avoid they must have made lots of mistakes, some with more disastrous consequences than others.

Biologists consider that many of the exotic compounds that plants synthesize are toxins and have evolved as grazing deterrents or as a form of chemical warfare in the ongoing battle against diseases. Others argue that many of the more bizarre chemicals are non-adaptive and purely the by-product of the complex metabolic reactions required for life. Grasses have taken a different approach. They are able to cope with the chomping of herbivores because they hide their tender growing points deep within the turf. In addition, grass leaves tend to be defended by physical rather than chemical defences. The edges of blades of grass cut through fingers like razor blades, because they are armed with rows of miniscule blades made of silica. This lack of toxins potentially makes them more ideal candidates for domestication and we will return to them later.

If plants are so well armed with so many chemical defence mechanisms, how come they are still susceptible to diseases and consumed by herds of herbivores? The answer is of course obvious. In any long running battle, both combatants are forced to develop strategies to deal with the tactics employed by their opponents. The struggles between hosts and diseases or between hungry animals looking to find dinner and plants trying to avoid becoming dinner are both examples of this. This process has many parallels with arms races that have been observed throughout human history. New weapons have repeatedly been developed and deployed that have quickly been rendered obsolete by new defensive structures. Evolutionary biologists call this process the Red Queen effect, named after the colourful royal in 'Alice Through the Looking-Glass' who is forced to run ever faster just to stay in the same place. Similarly, plants that develop new defences rarely if ever escape being consumed by herbivores or avoid playing host to hordes of pathogens. Any associated grazing animals, parasites or disease-causing bugs are forced to evolve methods to disarm plant chemical defences or to risk extinction.

The ongoing battle between plants and their herbivores as consumed and consumer, has developed the Red Queen effect to an entirely new level of complexity. Many species of animal have not just adapted to be able to detoxify the plants' defences; they have evolved mechanisms that are now dependent on them. For example, many insect pests actually locate their host plants by homing in on the defence chemicals that initially evolved to deter them. There are many examples where insects actually 'confiscate' the plant's chemical defences and use them to protect themselves from predators. In very crude terms this ability drives the difference between moths and butterflies. Moths generally don't have the ability to confiscate (or sequester) plant defence chemicals. Thus they are typically palatable to predators, which they avoid by being cryptic and nocturnal. In contrast, many butterflies can utilize plant defence chemicals as a form of protection, indeed they actively seek out more toxic host plants. As a result butterflies tend to be unpalatable, brightly coloured and fly during the day. There are a few moths that have also mastered this trick. The caterpillars of cinnabar moths thrive on toxic ragwort and burnet moth caterpillars seek out birdsfoot trefoil plants that contain elevated cyanide levels. As adults both these moths have distinctive black and bright red warning colouration and fly during the day. It is not actually correct to describe such creatures as being addicted to plant toxins. However, these evolutionary skirmishes are so important and such permanent features of nature that biological interactions are impossible to understand without taking them into account. This is also true when we try to understand our own interactions with the plants we consume, particularly when we consider the plants that we have been drawn to precisely because they contain high levels of potentially harmful chemicals. In many of these cases domestication seems to have resulted in them becoming more toxic rather than more benign.

Wasabi toothpaste – surely not

Many of our more esoteric crops are highly and distinctively flavoured and this is why we cultivate them, rather than for sustenance. Of course, everybody knows that spices were originally used to mask the flavour of putrefied meat. This is why we associate spicy foods with hot countries where it was more difficult to store meat before the invention of refrigeration. While there may be some truth in this, the real interactions between spices and microbial spoilage are far more complex.

Unless you believe that God benevolently created plants with fiery flavours in tropical countries for the purpose of disguising the flavour of rotten food (rather than liberally supplying freezers) then the tropical distribution of spices demands explanation. There are in fact good biological reasons, which account for the observation. The compounds responsible for the complex aromas and tastes that make spices spicy are thought to have evolved as defences against microbial disease or to deter grazing animals. The highest levels of biological diversity are found in the tropics. Here plants are more likely to be challenged by disease and attacked by animals. In the hot steamy tropics, the battles between consumer and consumed are at their most intense, and as a result the armoury of protective chemicals deployed by plants are more extreme.

There are a handful of spices that originated in cooler climates. Most notable among these are three related plants, mustard, horseradish and wasabi. These are all relatives of the cabbage and derive their incredibly hot tastes from mustard oils that are common throughout the family. The active ingredients in mustard oils are isothiocyanates, which are produced as the result of a rapid chemical reaction that occurs when the tissues of these plants are damaged.

Wasabi grows wild along the cool mountain streams of Japan. Wasabi has been grown in the region for more than 1000 years; however, because it has very demanding habitat requirements it is not widely cultivated and remains effectively undomesticated. This fact tends to limit its availability and thus most wasabi consumed outside Japan is probably horseradish with added green food colouring. Horseradish is prepared by grating its fibrous roots, while wasabi condiment is made from its fleshy stems. Both of these activities are best performed outdoors as the volatile chemicals that are released cause a burning sensation in the eyes and nasal passages. This fact enabled a group of Japanese scientists to win the Ig Nobel prize for chemistry in 2011. Their contribution to science was to develop a wasabi based fire-alarm with the ability to wake sleeping deaf people within ten seconds of the device releasing isothiocyanates. Fortunately the smell disperses quickly, so that recipients don't have to deal with wasabi fumes as they head for the fire-escape. Conversely, if you are an aficionado of wasabi this means that you only have 15 minutes within which to fully enjoy it, before it loses its potency. Thus, real connoisseurs insist that it's made fresh on every occasion.

The fact that sushi chefs usually apply wasabi paste in its fresh state between layers of raw fish and rice rather supports the hypothesis that powerful spices are used to mask unpleasant flavours. But wasabi does more than just disguise the smell of fish; it has anti-microbial properties that have been demonstrated to be effective against *Vibrio parahaemolyticus*. This species of bacteria is commonly associated with seafood related food poisoning which is particularly common in Japan. The isothiocyanates found in wasabi have also been found to help control other bacteria which are associated with food poisoning including: *E. coli*, *Staphylococcus aureus* and *Helicobacter pylori* (which has been shown to cause gastric ulcers). Perhaps strangest of the lot is the fact that wasabi helps control *Streptococcus mutans,* the bacteria that causes tooth decay. This has led to a number of scientists suggesting wasabi should be included in toothpaste. Although it might be as effective as fluoride, this suggestion sounds even less likely than a wasabi fire-alarm.

Chillies – some like it hot, some do not

Chillies are perhaps the spiciest of the spices and are among the oldest cultivated plants from the New World tropics. A team from the Smithsonian have recently claimed to have found evidence of chilli consumption in Ecuador from more than 6000 years ago. Today chillies are cherished around the globe and are used in 'traditional' recipes in

Africa, Asia, and Europe and from their home territory in the Americas and the Caribbean. Chillies are available in a dazzling array of different forms, from sizzling hot habañero and Tabasco peppers to the mildest of sweet pimentos; they vary in size and shape from small and thin bird-peppers to large and round bell-peppers, the fruit are reds, greens, yellows, golden to almost white, purple to virtually black. Actually this rainbow of variation is rather cheating, because it does not represent a single species. Chillies are one of a small group of crops in which several species are all grouped together as one. In the case of chillies, five related species have been domesticated on different occasions. The most widely grown is *Capsicum annuum* which originated in Mexico and the Southern United States. Moving south, there is the inappropriately named *C. chinense* and the very closely related *C. frutescens* from the Caribbean and northern South America; these give us the hot Tabasco and Caribbean Scotch Bonnets. In the heart of South America and the high Andes are another two species, but these are little cultivated outside the region.

Biologically the entire purpose of a fruit is to attract the attention of a hungry animal and then to be consumed and thus act as a dispersal mechanism for the seeds within. This is why fruit ripen to bright attractive colours and become sweet and delicious just as the seeds are ready to be dispersed. But this raises the obvious question: If that is true, why are some fruits poisonous or burning hot like the chilli? The answer to this riddle is: It rather depends on who you are trying to attract. If seeds are consumed by the wrong species of animal, instead of being conveniently dispersed to a suitable location with a little pile of fertilizer, they end up being digested along with the surrounding fleshy fruit. Here lies the true genius of the chilli. They simultaneously manage to attract the birds that disperse their seeds, while repelling pesky mammals that just eat them. Capsaicin, the chemical which gives chillies their fiery hotness specifically targets the nerve endings of mammals, causing exactly the same reaction in the brain as being physically burnt, but capsaicin is completely benign to birds. It is no surprise therefore to learn that birds are effective dispersal agents of chillies. In contrast chilli seeds are usually destroyed in the guts of the few mammals brave enough to risk being burnt.

Humans, as so often in life are just as perverse, and many of us are actively attracted to the burning hotness that is caused by eating chillies. To the extent that they search out ever hotter varieties as their nerves become desensitized to the capsaicin. Plant breeders compete to produce increasingly hotter varieties to supply these chilli junkies. In 1912 the chemist Wilbur Scoville invented a scale for describing the hotness of chillies. This involves dissolving a sample of chilli in sugar solution and diluting it until the point that the heat is no longer detectable by a panel of five volunteer tasters. The Scoville scale ranges from the sweetest bell-peppers at zero on the scale, to 16,000,000 for pure capsaicin. The mighty scorching habañeros are around 350,000 to 580,000, which are mild compared to the ludicrously nuclear naga chilli at 923,000. These are sold with extreme health warnings and are best handled only with gloves.

The Scoville scale actually demands a little more explanation, because capsaicin is insoluble in water but is soluble in alcohol. Therefore the sample of chilli to be tested must first be dissolved in a known volume of alcohol before being mixed with a sugar solution. This is said to be the reason that you are not recommended to drink water to extinguish the flames of the chilli incendiary device, but a beer might prove more effective. But then again without the beer you may never have turned arsonist in the first place and in reality, the alcohol content of beer is so low that it is not sufficient to effectively dissolve enough capsaicin to quench the fire. Even so, a hot chilli and a cold beer remains a classic combination.

Fig. 5.1 Chillies may be burning hot to human taste, but are not to the birds that are responsible for dispersing their seeds.

Saffron – a spice to die for

The word 'spice' is not only the term applied to aromatic plant products used in flavouring it is also suggestive of something exciting, of something exotic, and of something with that little bit extra. It is perhaps surprising therefore that about the only thing that most British people know about the world's most expensive spice, is that it used to be grown in Saffron Walden in Essex. In fact, apart from a few place names, history appears to have recorded very little else about the British connection with this most unusual and valuable crop.

The spice saffron is the dried stigmas of an autumn flowering mauve crocus. Only the few millimetres long receptive part of the female reproductive organ are utilized, although it is sometimes incorrectly said to be derived from the anthers, the male portion of the flower that produces pollen. The origins of saffron are ancient and obscure, but it is thought to originate in Asia Minor or the eastern Mediterranean. The saffron crocus is unknown as a wild species. Unlike the majority of life on earth whose cells contain two sets of chromosomes (one from each parent) the cells of saffron contain three. Although plants are much more tolerant than are animals of extra chromosomes, this extra set of chromosomes results in sterility and an inability to produce seeds. Saffron reproduces entirely by the vegetative splitting of its corms. This extra set of chromosomes may indicate that saffron was originally produced by the hybridization of two different species of crocus. Alternatively, species that tend to reproduce more frequently by vegetative means such as bulbs or corms, than by producing seeds are inclined to accumulate chromosomal abnormalities to the point where they actually lose the ability to produce viable seeds.

Not only is saffron the most expensive spice in the world, with the possible exception of the seeds of some orchids, it is also the most valuable of all plant products. Weight for weight gold is cheap in comparison. There are two reasons for this. Firstly only a minute fraction of the plant is harvested and secondly, the harvest itself is very labour intensive. To justify exorbitant prices, those in the trade delight in informing us how many thousands of crocus flowers are required to produce each handful of saffron. Yields vary considerably from more than ten kilograms of saffron per hectare under good conditions, to fewer than two in unfertile regions of Kashmir. This equates to between 70,000 and 300,000 flowers per kilo, or 180 to 750 hours involved in hand picking the flowers and pinching out their stigmas for drying. The annual world market in saffron is estimated to be based on ten billion handpicked crocus flowers. The majority of the flower is of no value and for approximately a month during the harvest huge heaps of discarded petals litter Spain, India and Iran, as they once clogged the streets of Saffron Walden.

The high value of the saffron crop has long made it an attractive target for adulteration, the simplest form of which was to allow the dried threads to absorb moisture and hence gain weight by storing them in a humid atmosphere. An alternative method for increasing the weight was by coating the saffron in oil, glycerine or honey. If the spice was sold in powdered form then other parts of the saffron flower such as the anthers may have been added, or indeed it may have been adulterated with other plant materials including turmeric, safflower or marigold. Such practices have always been frowned upon, to the extent that in medieval Germany, Jobst Findeker and his substandard saffron were burnt at the stake in Nuremberg in 1444 for their crimes. As an early form of spiced kebab!

There is a sad irony about Findeker's demise, because unlike today, where the use of saffron is almost exclusively culinary, in the medieval period it was used extensively as a medicine. It was thought to cure insomnia, colds and asthma, be effective against scarlet fever, cancers and smallpox, act as a sedative, a pain-killer and an aphrodisiac and be a diaphoretic, an emmenagogue and expectorant. In reality it is probably none of these things, however it is a neurotoxin, but only at doses likely to bankrupt before poisoning. All the same, Findeker was probably doing the citizens of Nuremberg a favour by diluting their dubious cure-all.

Even more poisonous is the autumn crocus or meadow saffron, which has sometimes been confused with true saffron. This crocus look-alike is actually a member of the lily family. Its pale purple flowers can be found growing wild in old damp pastures in southern Britain, but it has become increasingly rare because it has been destroyed by farmers trying to prevent the poisoning of cattle. The autumn crocus was formally used to treat gout and today is the source of colchicine, a powerful drug used by geneticists to prevent cells from dividing and to help visualize chromosomes.

Although there is little in the way of direct evidence, it seems likely that the saffron grown in Essex and Cambridgeshire during the fifteenth, sixteenth and seventeenth centuries was intended neither for medicinal nor for culinary use. In fact there appears never to have been a tradition of using saffron in the kitchens of Essex. Instead it was probably primarily used to produce a yellow dye. Before Saffron Walden was so called, it was an important market town trading in wool and yarns. Modern synthetic alternatives have long since replaced saffron as a dye. In the end therefore Findeker's successors triumphed, at least in part.

Fig. 5.2 Saffron is one of the most expensive crops and is worth considerably more than its weight in gold.

Parsley, sage, rosemary and thyme

Our spices primarily originate in tropical regions and their fiery natures are thought to have evolved as a defence against the ravages of biological diversity determined to consume them. In contrast, most herbs can trace their ancestry to the Mediterranean. The difference between herbs and spices is somewhat arbitrary. Generally herbs are derived from the leaves of lower growing (herbaceous) plants and are frequently used both fresh and dried. Spices are a much more diverse set of typically dried plant products that include; bark, flower buds, seeds, seed cases, roots and stems. A few plants such as coriander are used both as herbs and as a spice. What they do have in common is that both herbs and spices seem to have powerful anti-microbial properties and this has greatly shaped how we have consumed them. Since ancient times, herbs have been utilized in four different and sometimes overlapping ways. Today we primarily think of culinary herbs as being used to flavour foods. The fact that herbs, as with spices, also help prevent bacterial spoilage is now an interesting novelty, but before the invention of refrigeration, this ability may have been important in reducing the ever-present risk of food poisoning.

Historically we relied upon the anti-microbial properties of herbs to a much greater extent, not just in cooking. Herbs were the bed-rock of the medical profession, indeed many university botany departments originated from within schools of medicine, as the study of plants was primarily focused on their health giving properties. Medicinal herbs were prescribed for all sorts of ailments. Parsley for urinary tract infections and flatulence, sage was taken for memory problems and Alzheimer's, rosemary helped relieve muscle pains and restore hair growth, while thyme was applied to cure acne and thrush. Many such traditional remedies have been investigated and their active ingredients have been integrated into modern medicines. These compounds are often the volatile oils with anti-microbial properties that attracted human interest in the first place.

Until the advent of modern chemistry and plumbing, herbs were also used as air fresheners and as insect repellents. In medieval Britain, where bathing had become a rare occurrence, powerful smelling plants such as meadowsweet, camomile, tansy, hyssop and wormwood were frequently liberally strewn around the homes of both rich and poor alike. These strewing herbs not only helped to cover the odours of the unwashed inhabitants, but also acted as a form of pest control, helping to repel fleas, ticks and clothes moths. This tradition has all but died out, the last vestige being potpourri. This fragrant mix of herbs, spices and wood shaving is now generally found in middle class bathrooms, perhaps suggesting that the owners don't entirely trust the washing process to be effective.

The fourth way in which herbs have been used is as part of religious activity. Since ancient times, herbs have not only been seen as having medicinal properties, but also sacred ones. It is easy to see how the two could become linked. Parsley was associated with the devil because its seeds took so long to germinate. It was believed that they had to visit Hades before they would do so. Sage was considered to be a sacred plant. It was burnt to drive away evil spirits and believed to prevent ageing and death. In Roman times rosemary was associated with the goddesses Minerva and Venus. In Christian tradition the flowers of rosemary were said to have been turned from white to blue by touching the cloak of the Virgin Mary. Tradition has it that thyme was one of seven herbs placed in Christ's crib at the nativity and that rubbing the juice of thyme flowers on your eyelids will give you the power to see fairies.

The psychoactive diviner's sage, which is a close relative of sage is sold legally in many countries. It is native to the cloud forests of Mexico. Here, the Mazatec peoples considered the plant to be sacred and their shaman used it to induce visions and hallucinations as part of ritual healing ceremonies. Most of the plant's local names allude to the fact that it is considered as an incarnation of the Virgin Mary. Diviner's sage's mind-altering abilities have been attributed to a terpenoid chemical called salvinorin A. Thus, there remains a possibility that the ability to see fairies, induced by thyme flowers, could be attributed to the same aromatic compounds that give herbs their distinctive flavours, smells and antimicrobial properties. In other words, the utility of herbs as culinary products, medicines, air fresheners and in religion, may all relate to the fact that they are rich in volatile oils containing aromatic monoterpenes and sesquiterpenes.

The observation that humans have cultivated so many low growing Mediterranean species of plant because they contain high levels of volatile oils begs the obvious questions; why do herbs produce these unusual smelly compounds and why are they associated with hot dry summers and milder wetter winters? The answers to these questions are not instantly obvious and almost counterintuitive. Places with Mediterranean climates tend to be susceptible to regular fires during the dry summer months. The vegetation, which grows in these areas, therefore has to be adapted to cope with bush fires. Although ironically the volatile oils produced by these plants mean that they are more prone to ignite. The resulting fires are rapid burning and tend to pass through the vegetation quickly before causing more extreme damage. It also has been shown that volatile oils help to reduce fire damage to plant cells by protecting membranes through a mechanism that is not yet fully understood. An extreme example of this phenomenon is the biblical burning bush. It has been suggested that the flammable forage may have been a plant called *Dictammus albus* that actually produces benzene along with other volatile compounds. Alternatively, Professor Benny Shanon of the Hebrew University of Jerusalem has argued that Moses' vision of the burning bush was a hallucination induced by inhaling aromatic oils produced by *Peganum harmala*. Either explanation depends on the chemical properties commonly associated with plants of the region and of our herb gardens.

Fig 5.3 Thyme, like other herbs, contains high levels of volatile oils that tend to make them flammable.

Willows – a bitter pill to swallow

Humans have been consuming plants for their medicinal properties almost as long as they have been eating them for nutritional reasons. In fact, it is often meaningless to try to disentangle the two functions. As early as the fifth century BC, Hippocrates, the father of medicine was recommending taking a bitter extract of willow bark to ease aches and pains. How this ancient herbal remedy led to the discovery of aspirin is a tale complex enough to give you a headache, with various scientists staking a claim and two very different plants vying for the starring role.

It is a commonly held misconception that willow bark contains aspirin. This is simply not true. The pharmacologically active chemical in willow is salicylic acid, which derives its name from Salix the Latin name for willows. A less widely held tradition claims that the bitter taste of willow bark results from Christ being whipped with a willow wand as a child. In fact salicylic acid is not unique to the willow, but is widespread in the plant world, where it acts as a plant hormone, involved in stress tolerance and disease resistance. In a few plants such as willows and meadowsweet, salicylic acid is present in high enough concentrations to be medically active and to the extent that it is potentially toxic to species feeding on them. In Europe, robins feeding on the seeds of meadowsweet come close to taking a lethal dose, while in North America, beavers which eat excessive quantities of willow bark escape poisoning by perspiring salicylic acid.

Salicylic acid was first extracted from willow bark by Johann Büchner, professor of chemistry at the University of Munich in 1828. A year later, the French pharmacist Henri Leroux, extracted larger quantities of purer crystalline salicylic acid from meadowsweet. Although salicylic acid does relieve pain it is not an ideal medicine because it causes digestive problems, including gastric irritation and diarrhoea. Therefore, throughout the nineteenth century chemists struggled to develop a less irritating formulation for salicylic acid. By 1853 Charles Frederic Gerhardt had synthesized acetylsalicylic acid (the compound later named aspirin) and even marketed the product during the 1880s. However, the discovery of aspirin has been widely credited to a German, Felix Hoffmann working for the drug company Bayer in 1897. The story goes that Hoffmann was looking for something to relieve his father's rheumatic pains. Acetylsalicylic acid was marketed as aspirin from 1899, a name derived from the then Latin name for meadowsweet (*Spiraea*) and not willow! In 1949, shortly before his death, Arthur Eichengrun, another German chemist working for Bayer at the time, cast doubt on this story when he made a credible claim to have been responsible for directing the research at the time. Whatever the truth of the story in the laboratory, the fact remains that the naturally occurring chemical is as effective today as it was in the time of Hippocrates. So, if you are ever stuck in the Canadian wilderness with a stinking headache, you may gain relief by licking a sweaty beaver!

Willow trees have a long history of being used by man, but hardly one that qualifies them the status of a crop. Taxonomically willows are a nightmare, as their lack of morals means that virtually every species hybridizes with every other, from low growing creeping willows found in sand-dunes or mountain tops to magnificent stately trees. The resulting range of growth forms have provided humans with a wealth of materials that have been used for weaving into wickerwork, through to providing a strong lightweight timber favoured for making artificial limbs. Probably the most famous use of willow wood is in the production of cricket bats, without which English civilization would probably cease to function! Only rarely are plants defined by a single human use, but the cricket bat willow has such an honour.

In complete contrast the future of willow as a crop is far removed from the genteel English past-time of cricket. In the twenty-first century willow is being grown on a vast scale in enormous monoculture plantations across prime agricultural land. This industrialization of sustainable short-rotational coppicing is far removed from the flower-rich coppice woodlands of old. Willow's rapid growth and re-growth rates may be part of the solution to the world's growing demand for renewable energy, but the agricultural landscape associated with its cultivation would be very unfamiliar to Ratty and Mole and their willow wicker basket.

Tobacco – what a strange thing to do with a sacred gift

There are many human uses of crop plants associated with the drugs they contain that are just so strange and detrimental, that you are forced to wonder - how did anybody ever think of doing that? Tobacco is a great example of this. You may well ask the question - how did the Native Americans ever think of harvesting tobacco leaves, curing them, shredding them, putting them in a pipe, setting fire to them and finally inhaling the fumes? The question could be further refined to ask - exactly how many different species of plant did they try inhaling until they found one with the desired effect? Plus given what we now know about the harmful effects of tobacco on human health, why did they go to so much trouble? In fact the more you look at this question, the more difficult it becomes to answer.

There are 64 species of tobacco, which originally grew wild across the Americas, Australia and on a few Pacific Islands. Of these about ten species were used by Native Americans and one by Australian aborigines for religious or medicinal purposes. Today only four species are cultivated for consumption and of these just one (*Nicotiana tabacum*) makes up the bulk of the world crop. So what is the mystery surrounding the domestication of these species?

Nicotiana tabacum, like many crop plants is unknown in the wild, but again like many other crop plants its parent species are still found in the wild. There is however, something very strange about these parents. Tobacco is cultivated and consumed because it contains the toxic alkaloid nicotine, which acts as a stimulant and hallucinogen in the human brain. Nicotine is chemically synthesized in the roots and then transported within the plant to the leaves. However, in the wild parents of *Nicotiana tabacum,* once the nicotine arrives in the leaves, it is rapidly broken down to become ineffective and it seems reasonable to assume that the same was once also true of *Nicotiana tabacum.* Exactly how the Native Americans managed to produce the domesticated tobacco plant in which the nicotine survives in the leaves is completely unknown. It is unclear why they cultivated this species in the first place if it contained no nicotine in its leaves. The Native Americans turned to the Divine to answer this question. According to Huron legends, in the ancient time the land was barren and the people had nothing to eat, so the Great Spirit sent a woman to feed the people of the earth. As she walked through the land, anywhere her right hand touched the ground potatoes sprang up and wherever her left hand touched the ground corn grew. When her job was finished and the earth was fertile and productive, she sat down and rested, and from where she sat and rested sprouted the first tobacco plants.

Although the various tribes of Native Americans differed in their beliefs about the details of the origin of tobacco, it was widely held to be a gift from the Great Spirit, given to aid communications between humans and the spirit world. To this end, they tended not to use tobacco recreationally or routinely, but occasionally and in much higher doses for religious purposes. Shaman or medicine men would take high concentrations of tobacco frequently as an enema as part of religious ceremonies. In such high doses tobacco induces hallucinations and it was believed that this allowed the Shaman to communicate with the spirits. Alternatively, when smoking tobacco the exhaled smoke was believed to carry a person's thoughts to the spirit world. In contrast, because tobacco use was considered sacred, the routine recreational use of smoking was typically regarded as being disrespectful to the gods. Possibly being the equivalent of making a nuisance telephone call to god! Interestingly, such disrespectful behaviour was considered likely to be punished by poor health. In contrast the traditional Native American method of consuming tobacco, (infrequently and in high dosage) is thought to be much less damaging to health than is the regular addictive smoking of cigarettes. But then again a shaman with a 30 a day habit might find sitting in a doctor's waiting room rather uncomfortable.

Cannabis will make you free

Hemp or *Cannabis sativa* is probably the oldest non-food crop of all. It has been cultivated in Asia to produce ropes, fabrics and oil-seed since Neolithic times. Its use spread as far as Europe by 1500 BC, but took another 2000 years before it became a significant fibre crop on the continent.

The narcotic properties of the plant have been enjoyed in India for over 3000 years, but perhaps because the drug content of the crop is dramatically reduced in cooler temperatures, this was not appreciated in Europe possibly until as late as the eighteenth century. Even so, for many years hemp was the most traded commodity on earth. The Great British Empire whose navy dominated the Seven Seas was dependent upon hemp canvas sails and hemp fibre ropes, (the word canvas being derived from cannabis). Without hemp Columbus would not have discovered the Americas, and American troops would have gone naked into battle during the Second World War. In spite of its great and glorious past, cannabis is now a minor crop on the world stage in terms of fibre, while as a recreational drug it is widely illegal and its possession is even punishable by death in some places.

Botanically, cannabis is in a rather small family along with the related hop plant. As with hops, cannabis is usually found as separate sexed plants, with the female of the species being favoured. Genetic control over gender in cannabis is highly complex, but with certain changes in day length and temperature it is possible for male plants to produce female flowers. The reverse sex change operation from female to male can be performed by a simple act of decapitation!

Female plants have been considered to produce the best quality fibre and to have more of a narcotic effect than do males and of course, because males do not produce seeds they are of little value in producing oil. For this reason over the years hermaphrodite plants have tended to be selected for, along with different varieties for fibre and recreational uses. The complexity of the different sexual forms of cannabis has resulted in many different terms being used to describe the various products derived from these different plants. Thus the words ganja and marijuana, although widely used to refer to the entire plant of either sex are more technically terms for the flowering tops of female plants. This has the potential to make life very interesting in a court of law, because since 1913, the cultivation, importation, possession or use of ganja has been illegal in Jamaica. The original use of the word ganja in Jamaican law was designed to differentiate its use as a drug from hemp grown for fibre. However, within the strict definition of ganja it would therefore not be illegal to cultivate, import, possess or use male cannabis plants. It is technically almost impossible to determine the gender of a fragment of hemp leaf material. This fact could be used to mount a spirited defence, as males are as common as females, so technically only half the crop is illegal. However, in spite of the reality of the widespread recreational use of ganja on the island and pressure from Rastafarians to legalize it on religious grounds, over the years the Jamaican laws on cannabis use have been tightened because of pressure from the United States. This happened first following the passing of the 1937 Marijuana Tax Act in the US. The newspaper magnate William Randolph Hearst was very active in lobbying for this legislation and campaigned against the evils of cannabis use. Perhaps his motives were not so innocent, because the imposing of restrictions on hemp cultivation and hence hemp paper manufacture greatly benefited Hearst's own financial interests in the timber pulp and newsprint industries.

There is another beautiful irony in America imposing its will on the legislation governing ganja use in the independent nation of Jamaica. In 1776 when Thomas Jefferson wrote the Declaration of American Independence, he did so on hemp paper. Furthermore this paper could have been grown on George Washington's hemp plantation and made in Benjamin Franklin's hemp paper mill.

Durian – the loathsome king of the fruits

No discussion about why we consume plants with unusual properties would be complete without consideration being given to the durian. The durian is a large tropical tree, a native of South East Asia that grows to a height of 25 to 50 metres. There are at least nine species of durian known to produce edible fruit, but only one *Durio zibethinusis* is grown commercially. Its fruit which can be as large as 30 centimetres long and weigh as much as three kilograms are born on the trunk and main branches of the tree rather than in its canopy. The phenomenon is called cauliflory and is not uncommon in tropical trees. It is thought to be associated with animal mediated pollination or seed dispersal. In Malaysia the large flowers of durians have been found to be exclusively pollinated by cave fruit bats. This unusual method of pollination could help explain why the durian is not widely grown outside the region.

The durian is infamous for its fruit, which appear to be loved or loathed in equal parts. There are many famous quotes which attempt to describe the flavour and smell of the durian. Perhaps the most widely cited description comes from the naturalist Alfred Russell Wallace: "This pulp is the eatable part, and its consistence and flavour are indescribable. A rich custard highly flavoured with almonds gives the best general idea of it, but there are occasional wafts of flavour that call to mind cream-cheese, onion-sauce, sherry-wine, and other incongruous dishes. Then there is a rich glutinous smoothness in the pulp which nothing else possesses, but which adds to its delicacy. It is neither acid nor sweet nor juicy; yet it wants neither of these qualities, for it is in itself perfect. It produces no nausea or other bad effect, and the more you eat of it the less you feel inclined to stop". Wallace was obviously a fan. Durian devotees regard it as the king of fruits. In fact many say it is the finest of all foods. There are however, many descriptions that the durian marketing board must find difficult to swallow. These include the American TV chef Anthony Bordain who said, "Your breath will smell as if you'd been French-kissing your dead grandmother", while the travel writer Richard Sterling described the odour of durian "as pig-shit, turpentine and onions, garnished with a gym sock".

Human opinions are obviously divided about the merits of the durian fruit. In contrast, it appears to be highly appealing to wild and domesticated animals alike. Elephants, orangutans, pigs, even tigers and civet cats are all known to be fond of durian. Many of the durian's vernacular names such as "civet cat tree" and "civet fruit" relate to this. Even its Latin name "*zibethinus*" is derived from the Indian civet, *Viverra zibetha*. It is however, not clear if any of these names related to the fact that civet cats like to eat durians or that the fruit smells somewhat cat-like.

Human durian fans often describe it as being an acquired taste. This appears to be the case with many of the more unusual and strongly flavoured foods that we eat. Perhaps this suggests that some of the explanation for why we consume crops crammed full of exotic foul smelling, burning or hallucinogenic chemicals lies within ourselves rather than within the plants. Some of the compounds involved are obviously addictive. But with others such as the durian, and the chilli our taste buds appear to become desensitized to their effects and we need to eat increasing amounts to have the same effect. On top of this biological explanation there may be issues of cultural identity and status at work. Eating things that quite simply hurt, may make us appear more masculine, or may make us feel part of a group. Finally, the Red Queen herself (whom was discussed at the beginning of this chapter) may be at play. The amounts of spices that are typically used in cooking may deter the growth of food spoiling microbes. Very few bacteria are killed at the curry level concentrations. What we may be witnessing is an evolutionary arms race in which these crops are engaged in a battle to defend themselves from pathogens and humans alike. The plants may be evolving or being intentionally selected by humans to increase their toxicity generation on generation. However, in response to this humans (along with other herbivores, pests and the microbes that all compete to consume our crops) are also adapting, evolving, becoming accustomed to, even becoming addicted to the chemical weapons that the plants initially evolved to deter us.

References

Most significant new sources in the order they were utilized in this chapter.

Purseglove, J.W., Brown, E.G., Green, C.L. and Robbins, S.R.J. (1981) *Spices.* Volume 1. London and New York: Longman.

Crawford, R.M.M. (2008) *Plants at the margin, ecological limits and climate change*. Cambridge: Cambridge University Press.

Rainsford, K.D. (ed.) (2004) *Aspirin and related drugs*. London and New York: Taylor and Francis / CRC Press.

Goodman, J. (1993/2005) *Tobacco in history: the cultures of dependence*. New York: Routledge.

Bocsa, I. and Karus, M. (1998) *The cultivation of hemp: botany, varieties, cultivation and harvesting*. Sebastopol, California: Hemptech.

Subhadrabandhu, S. and Ketsa, S. (2001) *Durian: king of tropical fruit.* Wallingford, UK: CAB International: Daphne Brasell Associates.

6

Accidents of history

Some crops seem just too good to be true. How on earth did our ancestors manage to develop bananas that don't contain seeds? Or hybridize unrelated species to produce totally novel crops? This chapter covers a number of crops that have been created by our habit of growing related plants together and thus enabling them to unintentionally cross-pollinate each other. We also discover that although mutations are incredibly rare in nature, if you grow enough plants generation after generation, this random process can hit the jackpot and be responsible for the creation of new crops.

Humans have been domesticating crops for around 10,000 years. But not until 1900 and the rediscovery of Gregor Mendel's experiments with breeding peas, did we have any real understanding of the laws of inheritance. In fact, it was as late as 1676 when plant anatomist, Nehemiah Grew addressed the Royal Society that we started to appreciate that plants actually indulge in sex. In other words, for the vast majority of agricultural history we have not really had much of a clue about what we have been doing. The process of domestication has been one of simply identifying the most desirable or just unusual plants and propagating them. This duplication has occurred either by seed or vegetative means, from a cutting or simply dividing the plant into two. Although the sexual production of seeds has the potential to generate lots of variation amongst the next generation in comparison to vegetative propagation, given enough time, both of these processes are capable of delivering the genetic variation required for crops to become differentiated from wild plants.

In truly natural systems, genetic isolation of populations in places like the Galapagos Islands has been important in allowing species to change rapidly and differentiate from their parental stock. Similar processes have occurred during crop domestication, as humans have transported their crops with them during their relentless colonization of the globe. Unaware of the importance of maintaining genetic diversity, our ancestors almost certainly moved small numbers of seeds with them as they settled in new areas. This unintentional process of repeated genetic sub-sampling resulted in the dilution of genetic variation that caused crops to drift from their ancestors. Thus, even without the vagaries of fashion, regional prejudices or local selection, potentially lots of different local crop varieties have been generated as crops have spread around the world alongside farmers.

While genetic isolation of populations has been highly important in allowing evolution to occur more rapidly in natural systems, the exact opposite has frequently also contributed to the generation of new crop species. We humans are a highly mobile species. During the process of colonizing the earth, we have by necessity transported our crops along with us. At the same time we have embraced the new and cultivated similar local plant species alongside our familiar crop plants. Unwittingly, in doing so, we have enabled our old crops to hybridize with their distant wild relatives. Sometimes such unplanned hybridization has occurred with truly wild relatives and sometimes this has happened in cultivation. Frequently familiar crop varieties have been grown alongside their newly encountered local alternatives. This accidental generation of hybrids has been incredibly important during the process of crop domestication. The process may be entirely natural, but it has only occurred because humans have repeatedly and inadvertently conducted massive genetic experiments by engaging in the activity of farming. Throughout our agricultural history, growers with no scientific understanding of genetics have harvested millions upon millions of plants. In doing so they stumbled across these extremely rare freaks of nature. Without our loving care most of these new, fortuitous hybrid crops would have failed to thrive and disappeared as rapidly as they appeared. Indeed, the majority of our crops are unable to survive in the wild, and only persist because farmers and growers know how to accommodate their fussy requirements. Yet still we regard them as being completely natural.

In this chapter we discover that many important crops have emerged by good fortune and historical accident, as humans have grown different species together, unwittingly encouraging sex between distant relatives. Serendipity is required if you are to win the lottery. It appears that during the process of domestication, agriculturalists have won the lottery multiple times. This is only possible if you purchase enormous numbers of tickets. Multiply the number of plants (tickets) per field by the number of harvests over the last 10,000 years and you start to understand why so many of our crops are the result of improbably strange and individually rare occurrences.

The strawberry, true love or sexual frustration?

What could be more traditionally English than a bowl of fresh strawberries and cream? Well, the answer is just about anything, because the garden strawberry that is so much part of the Wimbledon scene is less than 300 years old, of American parentage and French nationality. Furthermore, cultivated strawberries are only distantly related to the species that grows wild in Britain, having four times the number of chromosomes. This difference is not just a mathematical curiosity but was to have important consequences for the development of the modern fruit. Strawberries with low numbers of chromosomes tend to have hermaphrodite flowers containing both male and female parts, while those with higher numbers are usually like humans in having discrete sexes, with individuals being either male or female. Finally, the traditional way to serve fresh strawberries was with sugar and Claret rather than cream.

The origin of the garden strawberry has been portrayed as a romantic love story of boy strawberry meets girl strawberry, but the reality was probably more about sexual frustration. The tale begins in 1556 when the Virginian strawberry first reached Europe from North America. The Native Indians who gathered the fruit to flavour beverages and breads had apparently planted it throughout the woods and meadows of New England. The fruit are not that much larger than those of the British native species, but its introduction was heralded as a great advance because of its wondrous flavour.

Many of the initial attempts at growing the Virginian strawberry failed. Not until 1624 was its cultivation successfully mastered. The reason for this was that the wild plants of this species are virtually all single-sexed, being either entirely male or female. At the time it was not fully appreciated that the basic fact of life, the knowledge that it takes both males and females to produce offspring, also applies to the plant kingdom. It seems probable that in ignorance of this fact, the early strawberry growers discarded their apparently barren male plants in favour of the productive female plants. The result was of course disastrous. Not until someone was fortunate enough to stumble across a rare hermaphrodite plant, whose flowers contained both male and female functions, was successful fruit production ensured. The whole situation was probably confused even further, because in Europe there is a common plant that looks almost identical to the wild strawberry. This look-alike species of *Potentilla* is known as the barren strawberry, not because it is genuinely infertile but because it fails to produce strawberries. Thus the growers of the time would have been very familiar with the idea that not all strawberry plants are capable of bearing fruit. Botany books from the period include the two species alongside each other, while modern DNA based taxonomy reveals that although they may look very similar, they are not so closely related.

This is how the story remained until 1714 when the large fruited Chilean strawberry was introduced from South America. French naval officer Amedée Frézier brought five plants back from his travels to Versailles. Unfortunately all five of these plants were females and indeed may have been derived from runners of a single plant. No one appears to have remembered the lesson of the Virginian strawberry from a hundred years earlier. By this time it had been discovered that flowers are sex organs. Even so, these five plants were distributed to various gardens, where they remained as unproductive curiosities for 30 years. Eventually, after this long period of fruitless frustration, it became clear that by growing alternate rows of the Chilean and Virginian strawberries, the South American females could be pollinated and produce their large fruits. Then, in some unknown place at some unknown time, this system of strawberry cultivation gave rise to the modern garden strawberry or pineapple strawberry as it was known. The modern hybrid strawberry is therefore the product of a South American female parent that had been deprived of sex for more than 30 years while being exhibited as a curiosity and which was then forced into having sex with a sexual oddity of another species. Hardly a text book romance.

This hybridization event probably occurred on several occasions, but it was in 1766 that the Frenchman Antoine Nicolas Duchesne was the first to realize that this new strawberry possessed characteristics that are intermediate between the two American species. He was able to confirm his theory by performing his own crosses. Some of the resulting progeny produced large flavoursome fruits from self-fertile hermaphrodite flowers. He had successfully recreated the modern garden strawberry. However, the French Revolution cut his studies short and his theory was not to be accepted until the twentieth century.

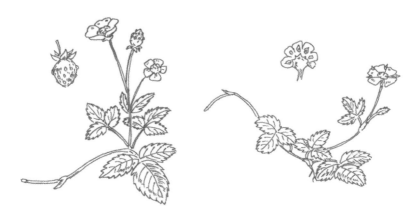

Fig. 6.1 Cultivated strawberries are not derived from their wild European relative. However, the confusion caused by barren strawberry, which looks like the wild strawberry but does not produce a fruit may have been an important part of the story.

Genetically modified wheat

Bread and the wheat from which it is derived are sold to us in down-to-earth language, as natural, plain and wholesome. Television advertisements employ pictures of bygone rural idylls, fields of golden corn ripening in the summer sun, with life simpler and slower, man and nature together in harmony. In reality this glorified grass, which by the hand of man has risen from obscurity to become the commonest plant on earth, is so out of touch with nature that it has been unable to survive in the wild for millennia. Outside cultivation this 'top of the crops' would be confined to history within a year or two.

Man has cultivated wheat for some 10,000 years, a practice recorded in the first book of the bible. However, far from being simple or natural, its genesis is so complex that modern genetic engineers can only marvel at the complexity. Unlike the genetically modified crops that concern us today, which contain just one or two genes from another species, bread wheat is an amazing three-way hybrid, assembled by combining all the genes from three different species of grass. To understand how this occurred requires an understanding of the mechanics of sex at a genetic rather than Karma Sutra level!

The fact that biologically sex is the combining together of cells from two individuals to form one new one hardly warrants explanation. A prerequisite to this sexual fusion is that the cells involved must first divide to reduce by a half the amount of genetic material they contain. The cells of most organisms contain two sets of chromosomes, one from each of their parents. Although shaped like sausages, chromosomes are roughly one million times smaller. They contain the DNA which carries genetic information. The cell division, which produces eggs, sperm, pollen or ovules, is technically termed meiosis. It involves the pairing of parental chromosomes, such that each chromosome from the male parent aligns with its matching partner chromosome from the female parent. This allows the cell to subsequently divide in half equally, to produce daughter cells containing a single copy of each chromosome, while facilitating the shuffling of parental genes. If this fails to happen then the offspring of every generation would contain twice as many genes as did their parents. Indeed this is known to have occurred during the domestication of some crops, but what happened in wheat is even more complex.

If two unrelated species have sex, then the hybrid offspring they produce are usually sterile. The reason for this is that in hybrids during meiosis chromosomes from the different parents are unable to find a partner with which to pair and thus the process aborts. Occasionally these abortive cell divisions do so in such a way that they produce viable cells containing duplicates of all the chromosomes present. The new doubled cells and their descendants are now fully fertile. Indeed they are technically, instantly a new species. This doubling provides each chromosome with an identical partner, with which to pair during meiosis. This enables fertility to be restored, but because they now contain as many chromosomes as both parent species put together, the new hybrid species is no longer able to interbreed with either of them. This miraculous hybridization method of creating a new species has occurred not once but twice during the evolution of bread wheat.

The origins of wheat have been traced to the biblical lands of the Fertile Crescent. In this region a primitive cereal called Einkorn wheat, which relies on just a single set of chromosomes, is still occasionally grown. Einkorn wheat was once thought to have been one of the three ancestral parents of modern wheat. Recent research points the finger of suspicion at a liaison between two distantly related wild grasses, with the Latin names of *Triticum urartu* and *Aegilops speltoides*. The hybrid species born from this pairing is cultivated in drier areas of the world such as India, the Mediterranean, and in parts of North and South America. It is used in making pasta and called unimaginatively pasta wheat, Durum wheat or in Latin, *Triticum turgidum*. Pasta wheat with its two sets of chromosomes was probably cultivated for many centuries before it also succumbed to the attractions of sex with one of its wilder relatives. On this occasion the act of adultery was with goat grass or *Triticum tauschii*, which still grows as a troublesome weed in and amongst the cereal fields of the Middle East. The progeny of this match contain six sets of chromosomes, two from each of its three parent species. Other hybrid species of wheat produced from other crosses are also known in cultivation, but none are as widely grown as this *Triticum aestivum* or bread wheat, which occurs as several thousand varieties. Always quick to learn from nature, man has been using the same doubled hybrid method since the 1930s to continue the process further. Both pasta wheat and bread wheat have been artificially crossed with rye to produce an entirely new and vigorous cereal called triticale which has either six or eight sets of chromosomes.

Compared to the laboratory techniques of modern genetic engineering, the hybridization events that gave rise to good old-fashioned bread wheat are natural, in that they were not deliberately planned or manufactured. But there is no escaping the fact that it is genetic modification on a massive scale. The rare 'natural' chance events involved only occurred because mankind unwittingly loaded the odds in their favour by cultivating millions of wheat plants over thousands of years and then carefully selecting and tending generations of plants incapable of unassisted life.

Does the wheat story have any relevance to current concerns about genetically modified foods? The answer is probably yes, in that one of the genes that was incorporated into bread wheat from its goat grass parent, codes for gluten production. This piece of genetic material enables wheat to produce large amounts of protein in its grains. This gluten protein is essential in the bread making process, because by trapping bubbles of carbon dioxide produced by the baker's yeast, it causes the dough to rise. Wash the starch out of a lump of uncooked dough by holding it under a running tap and you are soon left with just stringy elastic protein fibres. The gluten gene is therefore an important gene without which our daily bread would be hard and heavy. Unfortunately a minority of the human population is allergic to gluten, these sufferers experience irritation of the intestines, resulting in chronic diarrhoea. If a molecular geneticist had been responsible for incorporating a gene into an important basic food crop, which had such dramatic unpleasant consequences, the project would be stopped immediately. In addition to its impacts on human health, wheat has had even more dramatic effects on our environment. Being the most important crop on earth, millions of hectares of land have been put under the plough for its cultivation. Besides which, by feeding the heaving mass of humanity it has facilitated even more damage to the global ecosystem. The impacts of this particular genetically modified crop are almost unimaginable. Perhaps it is time to start a campaign to ban wheat.

Fig. 6.2 Wheat, which has played a pivotal role in the development of western civilization was produced as the result of a series of lucky accidents.

One banana, two banana, three banana more

The popular press often delights in informing us that the bureaucrats of the European Union wish to regulate for a straight banana. Within the truth behind such stories lies the irony that is the banana. It is certainly the case that the bananas found on most of the supermarket shelves of Europe vary in little more than price. In absolute contrast, the bananas and plantains cultivated throughout the tropics are so wonderfully variable that the species concept is unable to cope, and attempts to ascribe a single Latin name are a futile nonsense.

The name banana is loosely applied to cover several true species as well as a complex set of hybrids between them. Generally the term banana is used to refer to plants grown to be eaten raw, while the term plantain describes a larger and more angular fruit, which is eaten cooked. However, bananas can be cooked and plantains can be eaten raw and the two terms can be locally interchangeable. Additionally these plants have been cultivated not only for their fruit, but also for their fibres, leaves, stems and sap, as well as for ornamental purposes and for the shade they provide for other plants.

The majority of edible bananas and plantains are descended from just two wild species; *Musa acuminata* and *Musa balbisiana*, unfortunately neither of these really have common names. The first bananas to be cultivated were probably plants of *Musa acuminata* growing in the humid forests of Malaysia. In the wild, most of these plants produce fruits full of large seeds, but if unpollinated (and pollination is usually by bats) some plants develop fruit without seeds. Mankind, by choosing plants that are female sterile which do not produce seeds but yet still develop fruit, created the first edible bananas. Banana cultivation then spread north into the seasonally drier monsoon areas, where it came into contact with *Musa balbisiana* and because very occasionally bananas can produce seeds these two species were able to hybridize for the first time.

Most animal species contain just two sets of chromosomes (one from either parent) and are referred to as diploids. Plants, however, tend to be more tolerant of extra copies, this is particularly true of bananas. The outcome of what must have been the promiscuous sexual activity between the descendants of *Musa acuminata* and *Musa balbisiana* has been to produce plants with a whole array of chromosome combinations. If we use the letter A to represent a set of chromosomes from *Musa acuminata* and B for a set from *Musa balbisiana*, then the plants currently in cultivation include: AA, AB, AAA, AAB, and ABB. Plants with higher numbers of chromosomes such as AAAA, AAAB, AABB, and ABBB also frequently occur. Generally speaking, fruit commonly thought of as bananas contain only sets of A's while plants considered as plantains contain at least one set of B's. Plants with few sets of chromosomes tend to produce smaller, thinner-skinned fruit than those with multiple chromosomes. Within this vast cloud of variation, which defines most bananas, there are: fat ones and thin ones, straight ones and bent ones, sweet ones and starchy ones, red ones and green ones. And also something you would recognize on your supermarket shelf.

Meanwhile, outside this complex swarm of hybrid bananas, lie three further species of cultivated banana. The Manila hemp is from the Philippines and as its name suggests, is grown for its fibres. It has been used in the manufacture of strong ropes and also for less strong teabags. The Abyssinian banana from the highlands of Ethiopia is also grown for fibre, but additionally its starchy stems are regularly eaten. The cultivation of both these species is thought to be ancient but it is uncertain, with no direct evidence, that they were originally cultivated in the areas in which they now occur. Finally, the Polynesian Fehi bananas with their copper coloured thick-skinned fruit have been cultivated, not only because they are good to eat either roasted or boiled, but also for their sap, which yields a red ink.

The harvesting of bananas can be a messy business since the sap of all bananas causes indelible stains on clothes. Even so humans have used them in many different ways. In Central America, the sap of red bananas is collected as an aphrodisiac, by cutting the stems close to the ground and hollowing out the top of the stump. It is perhaps surprising that the banana does not have many more such sexual associations. Hindus are said to regard the plant as a fertility symbol. However, and perhaps naively, this has been attributed to the fact that the plant fruits throughout the year.

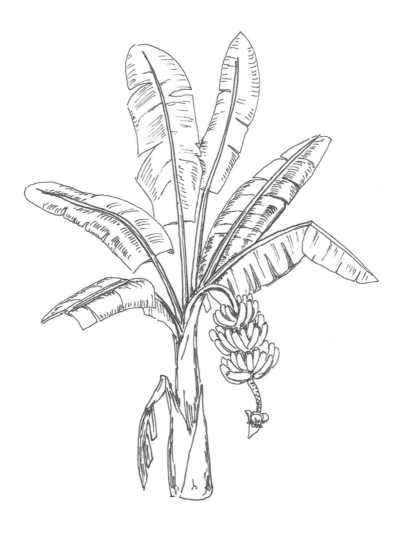

Fig. 6.3 Many bananas are male sterile, but maintain female fertility. If pollination occurs with a wild relative, their fruits may contain large seeds.

Citrus and the forbidden fruit of Barbados

Everyone thinks they can identify an orange or a lemon. However, the shrubby tropical and sub-tropical trees that we call Citrus have been cultivated for so long that after thousands of years of hybridization and selection their division into meaningful species is not as easy as you may think. This kind of phenomenon is not unusual in crop plants, but with Citrus mankind appears to have made this kind of change even more difficult to achieve by selecting for seedless varieties and asexuality.

Citrus are not unique in the plant world in that frequently when they produce what are apparently perfectly normal seeds, they are not the product of sexual reproduction but are in fact genetically identical to the mother plant. They are clones. Indeed, one of the reasons why the pioneering genetic studies of Gregor Mendel were overlooked for so many years was because of his failure to repeat his observations of the laws of inheritance. By unlucky chance his second species was one that produced its seeds asexually. Citrus do more than just produce seeds asexually; they also produce seeds with multiple embryos. Thus it is possible to grow several plants from a single orange pip. In fact within an orange pip it is often possible to find both asexual copies of the mother tree and sexual embryos; the true offspring of the same tree.

Within Citrus fruits there is a trend that the more intensively a species has been cultivated, the more likely it is to produce seeds via asexual means. Thus, oranges, grapefruit and mandarins are highly asexual, lemons and limes are partly so and citrons and pummelos reproduce via sexually produced seeds. This relationship may well have a causal basis, in that part of the 'domestication' process of any crop involves the selection for superior types. Once this has been achieved, then growers are interested in selecting types which breed true, and what breed is more true to type than an asexual species? But there is a problem here. All this occurred years before Mendel, when mankind had no understanding of the mechanics of genetics. Selection for asexuality is all well and good, but it does rather limit the scope for future hybridization and plant breeding.

Even with such a high degree of asexuality the Citrus family has still managed to produce an embarrassing mix of individuals of uncertain parentage. Take the grapefruit for example. Until Columbus' second voyage to the new world the Citrus clan was entirely unknown in the Americas. However, the grapefruit appears to have spontaneously arisen on the island of Barbados, being first reported in 1750. But from where did it appear? Opinions have been divided. It may have resulted from a chance mutation from the pummelo (also called pomelo and the shaddock after Captain Shaddock who introduced this large fruited Citrus to Barbados, while the French know it as pamplemousse which sounds more like something from a child's fairy tale). Alternatively the grapefruit may have arisen as the love child of an illicit liaison between an orange and a pummelo during those hot Caribbean nights! Perhaps this is why it was originally called the 'forbidden fruit of Barbados'. Whatever its origins, the grapefruit has been unable to shake off both these habits. It has subsequently given rise to pink fruited grapefruit apparently, by random mutations; or more technically by a transposon (a gene that changes its location within the DNA of a species). As a result of these 'jumping genes' branches of yellow-fruited grapefruit trees have been known to miraculously start to produce fruit with ruby coloured flesh. Furthermore, the grapefruit and the mandarin orange have sired a love child all of their own, the unromantically named ugli fruit.

What a load of rhubarb

The word rhubarb has been taken into common usage meaning: nonsense or rubbish. This is said to be related to the practice of actors silently mouthing the word to simulate background conversation. As such the word rhubarb could be used to describe the confusion surrounding the introduction and usage of the species in Western Europe.

The standard story of the introduction of rhubarb is quite simple. A plant known as rhubarb has been used by European herbalists since ancient times and dates from at least the Greeks, who probably imported dried roots from southern Russia or China. Its use as a laxative in Chinese medicine can be traced to 2700 BC. By the sixteenth century early European pioneers in South America were starting to return, not with the fabled gold of El Dorado, but with syphilis. The race was then on to introduce and cultivate rhubarb, as an infusion of its roots was thought to cure both syphilis and gonorrhoea. Marco Polo had hyped rhubarb to such an extent that the trade in dried rhubarb roots rivalled that in spice and opium in importance. Peter the Great set up a state monopoly in rhubarb and the Chinese prohibited the export of its seeds.

At this point in the story, 1573, we are told that a mistake was made and instead of introducing the supposed medically active 'Chinese rhubarb', *Rheum palmatum*, plants of 'garden rhubarb' *Rheum rhaponticum* were introduced. This impostor is what we now recognize as something eaten with custard, but at the time it was cultivated only for its roots. Quite quickly it was realized that this new rhubarb was unable to live up to its reputation, and the Herbalist John Gerard, refers to it as 'bastard rhubarb'. It is easy to imagine that you would not be overly impressed when your long-awaited cure for syphilis failed! Even so, it was almost two hundred years later in 1763 that the real 'Chinese rhubarb' was eventually introduced from Russia. The cultivation of Chinese rhubarb in Britain was encouraged by the Society for the Encouragement of Art, Manufactures and Commerce, which awarded prizes for those establishing large numbers of plants. However, it appears that the bottom fell out of the British laxative market and the cultivation of Chinese rhubarb never really took off. In contrast, the French discovered that the leaf stalks of garden rhubarb were delicious to eat and forced rhubarb stems started to appear in the fruit markets of London around the start of the nineteenth century.

The trouble with this story is that it is contradicted by many bits of evidence. In Culpeper's Herbal, written in about 1640 (some time after the supposed introduction of garden rhubarb, but before the introduction of Chinese rhubarb) both these species are described as being cultivated in Britain and as being as good as any imported. Furthermore he describes garden rhubarb, as culinary or tart rhubarb, implying that it was already being widely eaten. There are also curious records of an apothecary called Hayward from Banbury in Oxfordshire who in about 1777 appears to have cultivated and sold garden rhubarb from seeds sent from Russia in 1762. This is suspiciously close to the supposed date for the introduction of Chinese rhubarb. Hayward's rhubarb was said to produce a drug of excellent quality, which was peddled as the genuine rhubarb by men dressed up as Turks. It appears this may have been a confidence trick to pass garden rhubarb off as Chinese rhubarb, which was also known at the time as Turkey rhubarb. Whatever the truth behind Hayward's rhubarb, we are told that upon his death his plantations were left to his descendants and that these fields of rhubarb remain in cultivation around Banbury to this day. Thus, if the good people of Banbury are more regular in their habits than the rest of us, then perhaps the old apothecary was not being economic with the truth after all.

Further confusion is added to the tale by the fact that when garden rhubarb is cultivated from seed, it gives rise to plants which vary widely from each other. This strongly suggests that it is not a true wild species at all, but of some unknown hybrid origin. In which case, where did it come from? One final twist in the story is that in the French version of the saga, they are adamant that it was the English who were the first to discover that rhubarb was edible. Like much else about the rhubarb story, perhaps this inconsistency is more to do with marketing than with the truth.

One of the few things that are widely known about garden rhubarb is that its leaves are poisonous. Indeed many people are so suspicious of the leaves that they even refuse to put them on their compost heaps. Although it is true that rhubarb leaves do contain toxic oxalic acid which can cause swelling of the throat and tongue, the concentration is sufficiently low that it would take five kilograms of rhubarb leaves to be fatal. Thus, if you are thinking of slipping a little poison in the wife's tea, rhubarb leaves are probably not what you are looking for. While rhubarb may not be the ideal poison, it is a great illustration of the role of accidental introduction, hybridization and serendipity in the origins of our crops. Like many other cultivated plants, it is unknown in the wild and it is difficult to imagine that the first human who planted ancestral rhubarb had any clue where their action was to eventually lead.

Throughout the process of domestication, random chance mixed with human ignorance of genetics appears to have played an important role in the creation of our crops. This is true of minor crops such as rhubarb, through to the mainstays of our survival such as wheat. These are the real Frankenstein foods. Traditional agriculture has not just tamed the wild, but created plants that are totally unknown in nature. Many of our staple food plants are so far removed from natural systems that they would be extinct in a single generation if it was not for the hard work of farmers and growers, who harvest and sow seeds which have long since lost the ability to disperse themselves. Similarly agriculturalists use all the means available to defend their crops from pests and diseases, because many crops have been rendered defenceless by our dislike of their bitter taste. Interestingly, our ability to detect bitter chemicals in the first place is probably the result of natural selection, giving us the tools to detect alkaloids and avoid poisonous plants.

References

Most significant new sources in the order they were utilized in this chapter.

Readman, J. and Hegarty, P. (1996) *Fruity stories: all about growing, storing and eating fruit.* London: Boxtree Ltd.

Brown, T.A., Jones, M.K., Powell, W. and Allaby, R.G. (2009) The complex origins of domesticated crops in the Fertile Crescent. *Trends in Ecology & Evolution,* 24: 103-109.

Heslop-Harrison, J.S. and Schwarzacher, T. (2007) Domestication, genomics and the future for banana. *Annals of botany,* 100: 1073-1084.

Morton, J.F. (1987) *Fruits of warm climates.* Winterville, NC: Julia F. Morton.

Grieve, M. (1971) *A modern herbal,* Volume 1 & 2. USA: Dover Publications Inc.

7

Classic combinations and recurring themes

We have seen that the chance of a wild plant being domesticated is rather rare. However, a small number of plant families have independently provided us with important crops on several occasions. This chapter tries to identify what makes these families so special. It appears that each of these families produces seeds or fruits that are easily stored, or leaves that can be harvested over a prolonged season. In addition to this, it turns out that their nutritional properties and agricultural requirements perfectly complement each other.

It is said that humans consume every part of the pig, except for its squeak. Although the hog may be a versatile creature, the ways in which we utilize plants are far more varied. In spite of the fact that we routinely eat so few of the plant species that are available to us, we have found a seemingly endless list of ingenious uses for every plant product on offer. We dig up roots and underground storage organs, while fruits and seeds are devoured with relish. Plant sap is tapped to turn into rubber, or poured over pancakes as maple syrup or fermented to make birch sap wine. More viscous plant secretions give us varnish, glues and violin resin. Unopened flower buds provide us with the very different flavours of cloves and capers plus the rather less exotic tasting cauliflower. The stigmas and styles of crocus flowers are harvested to give us the spice saffron, and nectar is plundered from agave flowers to produce a range of desserts. The bark of trees provides us with items as diverse as cork, cinnamon and materials for building canoes. We use fibres from cotton and linen to make fabrics that are dyed using plant pigments. The fluffy fibres that surround the seeds of the kapok tree fill our pillows and stuff our teddy bears. The pharmaceutical and perfume industries both exploit the biochemical diversity of the plant kingdom to keep us fit and healthy, while smelling divine. Increasingly we harvest oil crops to manufacture plastics, drive our cars or fuel our energy demanding lifestyles. Over centuries we have learnt to tell the wood from the trees. So that the wood of the spindle tree is used for making spindles to spin wool, while boxwood is used for... well you get the point!

The variety of ways in which we exploit plants is simply staggering. At a simplistic level, the food plants that we cultivate can be classified crudely as those which have simple flowers that are easy to pollinate, while providing us with essential nutrition or interesting flavours. These are not difficult criteria to satisfy, and so our diets regularly contain a wide range of diverse plants. Yet of the 620 plant families that are available globally, four stand out because they have been the targets of domestication over and over again. Each of the primary Vavilovian centres of crop domestication, which are associated with early agrarian civilizations, appear to have completely independently selected these same four plant families. But what are these families, and what makes them attractive as potential sources of food?

Following is a list of the six primary Vavilovian centres of crop domestication; associated with each of these are their locally domesticated crops in each of the following four plant families; the cereals, the legumes, the cucurbits or melons and finally the Amaranthaceae or spinach family. Each of these ancient agricultural civilizations has developed traditional local recipes that incorporate their endemic crops.

Mexico and Central America

Cereals: Maize.

Legumes: Runner bean, small seeded common bean, lima bean and tepary bean.

Cucurbits: Squash, chayote, marrow, courgette and green striped cushaw.

Amaranthaceae: Quinoa, huauzontle and quelite.

Classic dishes: Frijoles Refritos (refried bean) and tortillas, Frijoles de la Olla and Tortillas de Maiz (stewed beans and corn tortillas) Black Bean Corn Casserole.

South America
   Cereals: Maize.
   Legumes: Lima bean, large seeded common bean and peanut.
   Cucurbits: Winter pumpkin and fig-leaved gourd.
   Amaranthaceae: Quinoa and qañiwa or canahua.
Classic dishes: The national dish of Argentina, Bolivia, Chile and
Ecuador is a thick stew called Locro, made of corn, beans and gourds.
There are many local variants which may include: potatoes, various meats
and chillies.

Mediterranean and Middle East
   Cereals: Wheat, oats, barley and rye.
   Legumes: Pea, broad bean, lentil and lupin.
   Cucurbits: Melon.
   Amaranthaceae: Beetroot/ sugar beet/ mangelwurzel, chard and
   goosefoot.
Classic dishes: Hummus and pitta, Chobra frik (Algerian crushed wheat
and chickpea soup), Chakhchoukha (semolina flat bread and chickpea
stew).

Africa
   Cereals: Millet, sorghum and teff (*Eragrostis tef*), a cereal of the
   Ethiopian uplands which is also knows as Williams Lovegrass.
   Legumes: Cowpea, Bambara bean, hyacinth bean and locust bean.
   Cucurbits: Muskmelon, white flowered gourd, oysternut and
   watermelon.
   Amaranthaceae: Doodo (*Amaranthus dubius*) and celosia.
Classic dishes: Food varies across the continent of Africa, but the
importance of these staples ensures there are some recurring themes. In
Nigeria, Moin moin (steamed bean pudding) is eaten with Ogi (millet
porridge). This is not dissimilar to Shiro, an Ethiopian paste of bread-
beans or chickpeas which are eaten served on Injera (teff flat bread). The
Yoruba people of Nigeria consume Iru (fermented locust beans) with
Egusi soup which is made of roasted pumpkin seeds and celosia or
spinach leaves. In much of the south of Africa maize porridge which has
many local names such as Sadza, Pap and Putu is eaten with bean relish
called Cankalaka.

China
   Cereals: Rice and millet.
   Legumes: Soya bean, adzuki bean and velvet bean.

Cucurbits: Winter melon (*Benincasa hispida*) or Chinese preserving melon.

Amaranthaceae: Sand-rice (*Agriophyllum squarrosum*) and Chinese spinach (*Amaranthus tricolor*).

Classic dishes: Bean tempura, tofu and rice, and stir-fried bean shoots and rice.

India

Cereals: Rice.

Legumes: Chickpea, mung bean, pigeon pea and urad bean.

Cucurbits: Cucumber, bitter melon and luffa.

Amaranthaceae: Spinach.

Classic dishes: The vegetarian cuisine of India contains many combinations of these crops, either cooked or served together. Examples of this are: Bisi Bele Bath (hot lentil rice), Dal Puri Roti (flat bread containing split peas), Mujaddara (cooked lentils and groats).

The grain of civilization

Many people regard grasses as being rather dull. However, cereals are probably the most important crops of all. Indeed it is possible to argue that human civilizations could not have developed without them. These agricultural grass crops have frequently taken on deep religious significance. Prayers are offered to ensure the delivery of "our daily bread" and phrases such as 'the bread of life' and 'bread of heaven' all emphasize their importance. Around the world several unrelated grasses have been domesticated primarily for their grains. There are a number of factors that make grasses good candidates for domestication. Many grasses have annual life cycles, so they can be harvested and re-sown in newly ploughed ground each year. Unlike most plants grasses do not rely on chemical toxins to ward off pests, but deploy physical defence mechanisms in the form of abrasive silica barbed leaves. So there is little fear of being poisoned by a grass. The energy rich seeds of grasses are dry and hard and can be stored for prolonged periods of time or traded between areas with good harvests and those where the crops failed. Finally several grass crops, including maize, sorghum millet, sugarcane, and teff have evolved a more highly efficient form of photosynthesis termed C4. This modified metabolism allows them to be more productive than other C3 plants, particularly in drought or low nutrient conditions. C4 or C3 refers to the number of carbon atoms in the first molecule in the photosynthesis biochemical pathway. Only three per cent of known plant species have C4 photosynthesis and yet they are responsible for 30 per cent of the carbon fixed by land plants. Because of this C4 plants have a significant impact on our climate. Not until around six and a half million years ago, did C4 grasses become abundant and spread across the world's plains and savannas, encouraged by their ability to tolerate grazing from herds of large mammals. Following this, the superior C4 form of photosynthesis was responsible for a reduction in levels of atmospheric carbon dioxide and an associated drop in global temperatures.

Not only are grasses probably responsible for the rise of human civilizations around the globe and for modifying the climate, it appears that grasses have also greatly influenced the ecology of the oceans. The silicates that grasses use to defend themselves were originally not widely available in many regions, including the world's seas. Following the evolution of C4 metabolism by grasses, the amount of silicates being washed into rivers and finally out to sea increased dramatically. This in turn was associated with an increase in the numbers of single-celled phytoplankton called diatoms. These diatoms have a siliceous skeleton. Until the evolution of grasses on land, diatoms were relatively rare because they were limited by the low availability of silicates. Today diatoms are ubiquitous and are thought to be responsible for most of the photosynthesis that occurs on the earth. Along with the grasses, diatoms are important in reducing atmospheric carbon dioxide levels and reducing global temperatures. And some people still think grasses are dull!

We've beans everywhere man

The second most important family of crop plants has to be the legumes. There are more than 23,500 recorded species related to beans and peas which equates to approximately eight per cent of all known plants. As with the cereals, leguminous crops were independently domesticated by all the early agricultural civilizations. Wherever you go, they have already bean!

The broad bean was first cultivated in the Near East during the Neolithic period and was well known to the ancient Greeks and Romans. They offered bean cakes to their gods and used beans as a form of ballot paper in elections, with black beans signifying a no vote and white beans a yes vote. Broad beans were originally small and black in colour, while more modern varieties have larger, green or white seeds. The writings of the Greek scientist, Dioscorides, reveal that even at this early date an over indulgence in beans had been associated with flatulence, dulling of the mind and in extreme cases, sterility. One can only assume that they were careful not to anger the gods by dedicating too many bean cakes to Apollo.

Throughout European history, there are references to broad beans being considered as unsuitable cuisine for the aristocracy, because of their effect on polite company. A more serious affliction is the inherited allergy to broad beans called favism, which destroys red blood cells and causes anaemia, a condition which is particularly common in Iran and around the Mediterranean. However, this problem and the fact that many legumes are highly toxic have not been sufficient to deter humans from cultivating them.

French beans and butter beans were transported from Brazil to Africa in returning slave ships. From there the beans travelled north into Europe. The new arrivals were welcomed as the long wished for flatulence-free beans. Pope Clement VII presented some of these new wonder beans to his niece, Catherine de' Medici on the occasion of her wedding to the future King Henry II of France. It is not known if there was an incident at the reception, but the gift certainly failed to live up to expectations.

The problem with beans is that they contain large quantities of complex carbohydrates, lots of fibre and even toxic compounds, which in combination are not easily broken down by the human digestive system. These carbohydrates pass into the bowel where bacterial fermentation converts them into as much as two litres of gas per day. Over the years various methods have been developed to avoid this problem. The simplest solution is prolonged cooking during which the complex carbohydrates are broken into simple sugars. The Americans employ a high tech method; they swallow enzyme tablets, sold under trade names such as 'beano' which avoid embarrassment by chemically enhancing digestion. Researching this problem took Dr Colin Leakey from Cambridge to South America and the markets of Chile. There he found apparently identical beans being sold for vastly different prices. When he made enquiries about this disparity the market traders appeared embarrassed but with the aid of basic sign language informed Dr Leakey that more expensive beans were associated with lower emissions. Returning to the lab he was able to successfully hybridize these new finds to produce a flatulence free bean that could be grown in the UK. As a nation the British consume two million pounds worth of baked beans a day. The potential implications of Dr Leakey's research are, therefore, not to be sniffed at.

Beans and peas like other members of their family, the legumes, contain bacteria in nodules on their roots which help convert atmospheric nitrogen into proteins. This ability means that they can be grown in soils of low fertility, where they act as a natural fertilizer enhancing the growth of other crops such as cereals. For this reason, peas and beans are always at the heart of agricultural rotations, and are usually grown the year before high nutrient demanding crops. The ability to fix nitrogen also explains why legumes tend to contain more protein than other plants. These attributes are part of the reason why legumes have been domesticated so many times, with many different species of beans being grown as a source of non-animal protein. In addition (and similar to the cereal grains) dried beans and peas are easily stored, making them a great food store for unproductive seasons. The proteins that legumes contain are particularly rich in an essential amino acid called lysine. (Amino acids are the building blocks of proteins and essential amino acids are ones that humans don't have the ability to manufacture themselves). Thus to remain healthy it is 'essential' that we consume foods containing lysine, such as legumes. Although legumes are lysine rich, they tend to contain very low levels of another essential amino acid called methionine. However, the good news is that cereals are high in methionine. Therefore, as any vegetarian will tell you, the combination of cereals and legumes constitute the basis of a healthy, balanced, protein rich diet. For centuries classic vegetarian cuisine around the world has been built around this partnership of grains and pulses. The mainstay of the medieval peasants of Europe was pottage, a thick stew of cereals and beans. The modern equivalent has to be – baked beans on toast.

In Africa cow-peas were domesticated for both their dried seeds, known as black-eyed beans and for the fresh pods, called yard-long beans or asparagus beans. This is different from the asparagus pea, which is a native of the Mediterranean cultivated for its young pods. In Madagascar the four-angled bean was grown for its edible root tubers, its fresh pods, fresh or dried seeds, and even its blue flowers can be eaten in salads. India is the home of pigeon peas and the hyacinth bean, a beautiful climbing plant which is now grown as an ornamental in the southern United States as well as for its flattened seed pods. Very few of today's crops originate from North America. The potato bean is a rare exception, its root tubers were not only eaten by the indigenous Indians, they also sustained the Pilgrim Fathers through their first winter in the New World. The Far East has given the world mung beans and adzuki beans, which are usually eaten as bean sprouts. The most economically important bean of them all – the soybean was also first cultivated in China about 3000 years ago. From China the soybean spread to Japan and then with the assistance of an amazing feat of botanical espionage it was smuggled into Europe. In the seventeenth century Japan was a closed society. This protected its culture from outside influence by minimizing all contact with the rest of the world. Thus it was that Englebert Kaempfer, of the Dutch East India Company and physician to the Governor of the Island of Deshima was allowed into Tokyo on the one day in the year in which the city was open to foreigners. Englebert did not waste his chance, and after bribing the local guards, he managed to grab a few soybean plants on route to the capital. Within Japanese and East Asian cuisine, edamame (cooked green soybeans pods) remain a highly popular dish.

As we have already seen in chapter three, four closely related beans were first domesticated by the native peoples of Central and Southern America. Runner beans originate from the cool uplands, common or French beans are native to the warmer temperate zones, the Lima or butter bean were first grown in sub-tropical climes and the tepary bean was cultivated in semi-arid regions.

Fig. 7.1 The nitrogen fixing abilities of legumes such as peas make them ideal crops.

Pumpkins, giants among crops

The utility of the four plant families that have been so universally domesticated is reflected in the fact that they were all amongst the first crops that humans chose to cultivate. However, the archaeological evidence of ancient use of the third of these families is rather limited, because melons, cucumbers, marrows and pumpkins etc., are fleshy and tend to leave very little trace of their existence. The cucurbits (members of the melon family) make such versatile crops that not only have many different species of them been domesticated, but recent genetic analysis has confirmed that this has occurred in *Cucurbita pepo* on several different occasions across South America. These different events have given rise to pumpkins, squashes and marrows. Even more remarkable is the bottle gourd that is thought to have originated in Africa. More than 5000 years ago the ancient Egyptians were using bottle gourds as water containers. Prior to this it appears to have floated across the Atlantic to South America, where there is evidence of it being grown as early as 9000 BP, well before Columbus made the trip. On both sides of the Atlantic, many different shaped hard-shelled fruits were selected for use as bottles, drinking vessels, ladles, floats for fishing nets and even musical instruments. Only the immature bottle gourd fruit are actually edible because when ripe (like many wild members of the family) they become incredibly bitter.

It can be rather confusing discussing the different members of the melon family, because the same common names such as pumpkin, squash and melon are frequently used to refer to several different species. Generally speaking fruits with hard skins that can be stored are termed pumpkins, whereas soft skinned types are called squashes. The word gourd is even more ambiguously used and is applied to similar looking fruit from completely unrelated plants that are used ornamentally or as containers. Perhaps the most bizarre example of this is the penis gourd. In the highlands of New Guinea, the male indigenous peoples wear the dried shells of the bottle gourd to cover their genitals. However, other plants are also used for the purpose and are also called penis gourds. The most common alternative being the carnivorous pitcher plant *Nepenthes mirabilis* which is totally unrelated to the bottle gourd. Considering the fact that *Nepenthes mirabilis* contains protease enzymes that are able to digest any animal matter that falls into its trap, it's a brave man that wears one as a penis gourd! Although anthropologists rather predictably regard the penis gourd as a sexual display, the Papuans assert that they are only worn to cover themselves and they would not consider selecting one larger than is genuinely required for the purpose. This view is supported by the fact that penis gourds are generally rather narrower than you might imagine. In contrast, pumpkins are the largest of all our crops. The world record weight for a pumpkin is in excess of a tonne, which is probably excessive, even for the most exuberant of sexual displays!

As crops, the cucurbits are remarkably different from both cereals and legumes. But perhaps they did not start that way. It has been suggested that they were first grown for their large oil rich seeds. Only later were mutant plants with non-bitter fruits identified. The cucurbits also differ from cereals and legumes nutritionally, in that they provide us with very little other than a few vitamins and fibre. In these terms the lexicographer Samuel Johnson may have had a point when he said "a cucumber should be well sliced, and dressed with pepper and vinegar, and then thrown out, as good for nothing". However, in regions with prolonged dry periods the cucurbits can be invaluable in providing a store of fresh vegetable matter. In arid places such as the Kalahari Desert, which is home to the watermelon, these fruit are a great source of clean safe water.

Fig. 7.2 The bottle gourd is unusual in being independently domesticated on both sides of the Atlantic.

Popeye, spinach and super-veg

Long before marketing departments had thought of selling us anti-oxidant rich fruits as 'super-foods', there was spinach. This annual, green vegetable is a member of the last of the four crop families that have been universally domesticated. In many ways the spinach family make the ultimate crops. They appear to have evolved the more efficient C4 form of photosynthesis on 15 separate occasions. Many members of the family have long growing seasons, during which their leaves can withstand being harvested repeatedly. Nutritionally they contain all nine essential amino acids, making them the perfect protein sources and as Elzie Crisler Segar's cartoon character, Popeye the Sailor Man knew, spinach is an amazing source of iron. Or is it?

Popeye first appeared as a cartoon strip character in 1929. Initially he acquired his super human strength by rubbing himself with a magic hen, but this fowl habit was soon replaced by an addiction to spinach! Although it has sometimes been claimed that Popeye was an invention of the spinach industry, this is not the case, although several important spinach growing towns in the USA have built monuments to the cartoon sailor as a mark of their appreciation. There is another more extreme myth associated with Popeye's fondness for spinach. For many years it was thought that spinach was selected by Segar because of its incredibly high iron content. This was in turn based on chemical analysis carried out in the nineteenth century by Professor von Wolff, who accidentally placed a decimal point in the wrong place, and thus, overestimated the iron content of spinach by tenfold. More recently however, this story has been questioned by Mike Sutton of Nottingham Trent University. It appears that several other early estimates of the iron content of spinach were all dramatically higher than the true value. This was not because of mass multiple misplacing of decimal points, but related to iron contamination of the samples due to bad experimental practice. The French military also seem to have been impressed by spinach's high iron content, because during World War I, spinach juice was added to wine given to soldiers suffering from haemorrhages.

Before the European colonization of the Americas, quinoa, which is a member of the spinach family, was first cultivated by the Andean people some 4000 years ago. Although it is a close relative of fat hen, which is eaten as a green vegetable, quinoa is harvested primarily for its grain. In central and southern America, several members of the spinach family were domesticated as pseudocereal. These plants are not grasses, but their grains are consumed in a similar way, often being ground into flour. As

with other members of the spinach family these pseudocereals are highly
nutritious, containing all the essential amino acids. However, because they
are not able to fix nitrogen, their total protein levels are lower than those
found in legumes.

Fig. 7.3 Quinoa is a member of the spinach family that was domesticated
in the Andes 4000 years ago for its seeds.

Sisters three – corn, beans and squash

In North America, backyard gardeners are encouraged to embrace the environmental knowledge of the Native American Iroquois tribe by planting a "three sisters" plot of corn, beans and squash. The entire project is frequently surrounded in mythology, muck and magic. The Iroquois are said to have believed in the three spirits of corn, beans and squash, and that these three spirits love each other dearly and can only thrive together. Each of these sister spirits or De-o-ha-ko were considered precious gifts from the Great Spirit given to sustain both humans and each other. Or in horticultural terms; maize provides the structure for the beans to climb up, removing the need for poles. The beans fix nitrogen in the soil to the benefit of the squash and the corn, while the squash spreads along the ground acting as living mulch keeping the soil moist and free from competitive weeds.

It is rather fortuitous that these three American sisters get along so well because they are the product of a rather dysfunctional family. Squash is probably the oldest sibling and may be as much as 10,000 years old. There are many wild species of squash, pumpkins and gourds in Mexico and Guatemala. Although these are bitter tasting, they all contain edible seeds and were probably first cultivated for making bowls and spoons. The middle sister is probably maize, 'born' about 6000 years ago, again at the hands of the indigenous peoples of Mexico. Beans are the baby of the family and like squash are in fact many different species that have been cultivated for thousands of years in both Central and South America. The sisters' living together as one happy family is much more recent than this. The Iroquois have been cultivating the three crops together for about 700 years and the term 'three sisters' only dates from the nineteenth century.

Similar multi-cropping systems or companion planting methods are widespread around the small garden agricultural systems of the world. They make good ecological sense as the different crops are complementary, exploiting both the soil and sunlight in different ways, utilizing their different root and stem architectures and growth periods. Mixed plantings also reduce the likelihood of a pest or disease sweeping through a monoculture field and provide an insurance against environmental extremes. The down side of companion planting is that it is labour intensive. In the case of the three sisters, there are very complex growing instructions about ground preparation, inter-planting distances and different planting times for the different crops. For example, the corn needs planting before the beans so they have something to grow up. Then of course they require manual harvesting at different times so as not to damage the later maturing crops. This is all very well in a back-yard plot, but not very practical on the industrial scale of modern agriculture.

What do the Iroquois sister spirits of corn, beans and squash tell us about environmental harmony? The three sisters of course, are members of three of the four plant families that have been so frequently independently domesticated. They do indeed grow in harmony with each other and this may in part explain their global success as crops. Independently, they represent some of the most photosynthetically efficient plants on earth and they all either store well or crop over extended seasons. When eaten in combination, these crops provide us with almost the perfect diet, because they are rich in carbohydrates and oils, and they contain a balance of proteins, vitamins and fibre. In the most arid regions of the world they not only feed us, but offer a safe supply of water, plus a drinking vessel from which to drink. It really should be no surprise that these four plant families have been domesticated so many times by all the world's great civilizations.

References

Most significant new sources in the order they were utilized in this chapter.

Harris, D.R., and Hillman, G.C. (1989) *Foraging and farming: the evolution of plant exploitation*. London: Unwin Hyman Ltd.

Beerling, D. (2008) *The Emerald Planet: How plants changed Earth's history*. Oxford: Oxford University Press, Inc.

Heywood, V.H. (ed.) (1993) *Flowering plants of the world*. Second Edition. New York: Oxford University Press, Inc.

Bisognin, D.A. (2002) Origin and evolution of cultivated cucurbits. *Ciência Rural*, 32: 715-723.

Heiser, C.B. (1973) The penis gourds of New Guinea. *Annals of the Association of American Geographers*, 63: 312-318.

Bush, M.B., Hansen, B., Rodbell, D.T., Seltzer, G.O., Young, K.R., León, B. and Gosling, W.D. (2005) A 17,000-year history of Andean climate and vegetation change from Laguna de Chochos, Peru. *Journal of Quaternary Science*, 20: 703-714.

8

Ownership and theft

By weight, plant products are some of the most expensive commodities on earth. You might expect therefore that this would have driven us to domesticate many more crops than we have. However, the reverse is probably true. This chapter includes several examples where crops have become so valuable that this has fuelled economic self-interest in those involved in growing and trading in these crops. This in turn has driven them to steal, smuggle, outlaw and even destroy these plants, to an extreme that has been damaging to our crop genetic resources.

Everywhere you look there are plants. Without plants there can be no animals and certainly no humans. We don't just eat plants. While doing so; we sit on chairs made of plants, eat at tables made of plants, and live in homes built of plants. We clothe ourselves in plant fibres. They are used to make musical instruments and most of our great literature and art was produced on plant material, coloured with plant pigments. We ferment plants to produce alcohol. Chemicals derived from plants make us high and are also still the basis of most of our medicines. The list goes on and on. And yet, we still utilize a tiny proportion of the plants that are available to us. Not only are plants essential for most human activities, the crop plants that we exploit in so many diverse ways are the rare elite. This can make them incredibly valuable and the people who control their cultivation and trade exceedingly rich and powerful. It is no great surprise therefore that human history is bursting with stories of subterfuge, stealing and smuggling of crops plants. Breaking monopolies of supplies of crops frequently motivated the great journeys of discovery such as those of Marco Polo and Christopher Columbus. Throughout history, mass human migrations have occurred as crops have failed. Industrial level slavery was invented to provide the manpower to cultivate sugarcane and to pick cotton. Powerful companies, even countries have fought one another to control the market for spices and drugs.

It is easy to argue that the control of crop plants has been a significant factor which has influenced the course of human history. Conversely, our history, specifically those struggles to monopolize or break monopolies in the production and supply of crops have significantly impacted on the crops themselves. Attempts have been made to increase the prices of crops by reducing their availability, and in doing so diminishing their gene pool. Such actions have had long-term deleterious effects on the crops concerned and now appear crazy. But of course those involved were ignorant of the need to conserve crop genetic diversity and were purely motivated by greed. Similarly, naval blockades have been used to limit the supplies of crops during wars and for simple financial gain. This has been another significant driver in crop domestication, forcing the besieged population to seek alternative plants. All of these factors may seem like distant history, but yet their legacy persists to this day. The losers in these historic botanical power struggles now fervently protect their endemic crop germplasm. In contrast, the major powers in world politics drive through conventions on biodiversity that enshrine the protection of crop genetic diversity for the benefit of all mankind.

Captain Bligh and the breadfruit

The story of the mutiny on the Bounty and the aborted introduction of the breadfruit from Tahiti to Jamaica is one that is well known to lovers of classic cinema. However, the tale told rarely gives the breadfruit plant the starring role it deserves.

The breadfruit tree has been cultivated by man for so long that its exact origins are unclear. It has certainly been grown since ancient times from India and the Malayan archipelago across the islands of the Pacific. A member of the mulberry family, the breadfruit plant is a tree some 25 metres tall, with large dissected leaves of almost a metre long by half a metre in width. The most commonly grown type is seedless with round fruit measuring up to about 30 centimetres in diameter. These statistics may appear dull, but they are an integral part of the story of Captain Bligh and the mutiny.

By the first part of the eighteenth century, the breadfruit had gained an undeserved reputation with the colonial powers of Europe, as a crop that could feed their expanding slave based colonies in the Americas. Thus, when a series of famines struck Jamaica, the plantation owners began to petition King George III to have the breadfruit introduced from Tahiti to feed their starving slaves. Curiously, because of rivalries between colonial powers, the plantation owners appear to have missed the fact that the French had already introduced the breadfruit into their Caribbean colonies. Indeed, even before Bligh set out on his ill-fated voyage, it appears that the British had captured a French ship bound for Martinique and already returned to Jamaica with the bounty which included breadfruit plants! Still this fact was not about to deter the great Captain and thus five years later in 1787, he and his crew set-sail for Tahiti in search of breadfruit.

What the filmmakers fail to appreciate is the scale of the task Bligh and his small ship had set themselves. As stated above, the breadfruit in question is seedless, and thus could not easily have been transported as a convenient packet of seeds. No, the breadfruit is a fair sized tree with leaves almost a metre long. Travelling half the way around the world under sail it would not be possible to stow the plants below deck, as obviously they needed to be in the light. So, picture this, more than a thousand (yes, 1,015 - Bligh was not one to do things by half measures) trees in pots with massive leaves flapping everywhere. With the drying winds of the open ocean, it is no great surprise that they used lots of water and no surprise either that the crew became fed up with the damned things and threw them over-board.

Not easily deflected from his task, four years later Bligh set out once more with a new crew and a new ship, the Providence. Apparently learning nothing from his experience on the Bounty, he doubled the number of breadfruit plants on board to more than 2000. Indeed, Bligh, being a man of ambition, stopped at the West Coast of Africa on route to Jamaica and picked up a consignment of akees, another tree crop intended to feed the slaves of the West Indies. Some reports however say that Bligh was responsible for transporting the akee from Jamaica to the gardens at Kew.

Although it was rapidly established around the British Caribbean, the breadfruit never did substantially feed the starving slaves of the West Indies and to this day it is often still held in low regard throughout the region where it carries the stigma of being slave food. It appears to be cropped as much for its rubbery latex, as for food. Breadfruit trees throughout the Caribbean carry the scars of numerous cutlass blows inflicted to tap the latex. The sap is rolled into balls and placed in the branches of trees where parrots appear to be stupid enough to become ensnared in the sticky stuff.

The noble sugarcane

Over the course of history a few crops have become so important as to dominate economies and shape the development of entire regions. These crops tend not to be the staples of our everyday lives but those expensive luxuries that add that certain something extra. As such it is almost impossible to predict from their humble origins which species are likely to become the mega-stars of the crop world. One such species is sugarcane and that certain something is sucrose. Once its use was exclusive to the upper classes, but now sugar is the standard fix for that human craving for sweetness.

Sugarcane is a large tropical grass, originally from the humid forests of Papua New Guinea. There is an unbroken tradition of sugarcane cultivation in this region, which stretches back into the mists of time. The native Papuans grow garden canes in clearings in the forest as much for ornamental value as for the sweet juice they contain. These varieties are thick stemmed and brilliantly coloured in an unbelievable array of yellows, oranges, reds, greens, blues and black. They can also be variegated with either horizontal or vertical stripes, and actually look like something you might expect to find in a sweet-shop window. In addition to selecting something pleasing to the eye, the Papuans selected canes with higher sugar concentrations and lower fibre content, producing stems that were more enjoyable to chew upon as they walked through the forest. Even so, it must be said chewing sugarcane is still like sucking balsa wood that has been soaked in syrup.

These native garden canes are unknown in the wild and rapidly die out when their forest clearings are abandoned. For this and other reasons, garden canes are considered to be a different species from their truly wild relatives. These cultivated garden types or 'Noble' canes have the Latin name *Saccharum officinarum* and are thought to be derived from the domestication of *Saccharum robustum* which grows wild along the rivers of Papua New Guinea. Until this century much of the world's sugarcane crop was either of the original Noble type or of a variety known as 'Creole' which is a natural hybrid of it. However, modern crop breeders have developed what they term 'Nobilization' which is a process by which they cross Noble garden canes with either of three of their wild relatives to incorporate desirable genes for disease resistance. Then via a series of back-crosses with their Noble parent the breeders endeavour to regain the required high sugar concentrations. By the time the process of 'Nobilization' is complete, some breeders privately question if any genes from the wild canes remain.

The cultivation of sugarcane by Europeans did not become well established until the colonization of the New World. It was introduced to the Caribbean from the Canary Islands as early as Columbus' second voyage in 1493. In spite of Columbus' enthusiasm for the crop, the sugar industry in the West Indies was slow to develop and relied heavily on state aid and subsidies from the start. It also relied heavily on manual labour and thus fuelled the evil trade in humans from West Africa to the New World. Inevitably slave based economies are inefficient and throughout the seventeenth and eighteenth centuries the Caribbean producers were given financial encouragement by the colonial powers. This policy generally referred to as the bounty system resulted in 'King Sugar' dominating trade within the region and artificially inflating land prices.

By the nineteenth century times were changing and the world sugar market was becoming increasingly chaotic. This was driven by a number of linked factors. There was competition from the developing sugar beet industry. This was stimulated in part by British naval blockades of Napoleon's France, which by 1811 forced him to decree the compulsory growing of sugar beet and to establish research into sugar beet cultivation. Ironically, within a few years the blockades were over and France then imposed restrictions on sugar imports to protect its developing beet industry. Similar beet industries were developing in Germany and Britain. These were supported by Wilberforce and the Abolitionists, who were campaigning against the bounty system which helped to prop up the slave estates. In addition to pressures from home, sugarcane growers elsewhere in the world were demanding an end to the financial support given to West Indian sugar. The outcome of all this was the 1846 Sugar Duties Act which started the process of equalizing the duties on sugar from different origins. In 1874 Sugar Tax was abolished and the last two decades of the nineteenth century saw the European powers boosting beet exports by paying subsidies on every tonne. The cost of sugar on the market dropped to far below the cost of production either from cane or beet.

By the twentieth century sugar had stopped being a luxury item and became an everyday commodity. This had knock on effects for the cultivation of other crops. In 1831 more than seven hundred gooseberry varieties were known. All of these were sweet varieties, which were often not picked until November to maximize their sugar content. Once sugar became cheap, virtually of these were replaced by the summer fruiting tart varieties we know today. Thus, while it is a vast over simplification to say that the emancipation of the West Indian slaves changed the nature of the British gooseberry, it is true to say that the two events are not unlinked.

In the twenty-first century exciting new possibilities are opening up for this most noble of crops. A close relative of sugarcane, called *Miscanthus* or elephant grass is being developed as a biofuel crop that can be grown in temperate climates. The dried stems of this grass are burnt to generate electricity. As part of the crop breeding processes many related species of *Miscanthus* are being hybridized and this could involve crossing it with sugarcane, with which it happily shares genes. So don't be surprised if you start to see fields full of something that looks very much like sugarcane in places you might not have expected.

Cloves face sudden death

Biodiversity conservation is currently seen as an essential part of saving the planet. However, over centuries the succession of nations that have monopolized the world trade in cloves has deliberately and successfully decimated genetic diversity within the crop, with potentially disastrous consequences.

Cloves are the dried aromatic flower buds of a small evergreen tree, which is native to the islands of the Indonesian archipelago. It is said only to grow on tropical islands where it is possible to see the sea. It is related to a number of other smelly trees such as eucalyptus and pimento, the source of allspice. The cultivation of cloves in the Far East is ancient and the spice has been known in Europe since the end of the second century. During the Han period (220-206 BC) all court officials were required to hold cloves in their mouths when addressing the Emperor, to ensure that their breath was sweet. For more than one thousand years the Chinese managed to monopolize this valuable trade by concealing the source of their supply. Cloves were imported into China before being exported to India and Europe via the spice caravans. Cloves have been used medically, as a cure for languid indigestion and flatulence. Clove oil has weak anaesthetic powers that have been used to dull toothache. In Indonesia large amounts of shredded cloves are mixed with tobacco and smoked in 'kretek' cigarettes. Current consumption is in the region of 36 billion cigarettes per year and it is this use, which now dictates the value of cloves on the world market. However, it was the use of cloves in flavouring both sweet and savoury dishes that made them the most valuable of spices for two millennia.

By the sixteenth century the Portuguese finally discovered the Chinese secret. The source of the clove crop was five small volcanic islands in the north Moluccas that were to become known as the Spice Islands. For the next century the Portuguese dominated world trade in cloves, planting trees throughout the Moluccas islands. When the region fell into Dutch hands the trade was taken over by the Dutch East India Company. For ease of control they transferred the growing of the clove crop to Ambonia and a few other small islands in the south of the Moluccas. To protect their monopoly they destroyed trees outside these few islands and imposed severe penalties on anyone caught growing them in prohibited regions. In 1816 this resulted in what has been described as the most fragrant fire in the history of the world, as thousands of clove trees went up in smoke. The fumes could be detected hundreds of miles out to sea. This act enraged the native Moluccans who traditionally planted clove trees on the birth of their children and believed that the fortunes of these trees were linked with those of their children. In the bloody uprising that resulted, more than just clove trees were killed.

In spite of all these efforts the French managed to smuggle some trees out and establish their own plantations on the French controlled islands of Mauritius and la Réunion. It is said that during this process only a single tree survived to become the ancestor of the entire crop in la Réunion and from there the clove plantations of Madagascar. Early in the nineteenth century trees were taken from this source to Zanzibar off the East coast of Tanzania. Within fifty years Zanzibar had become the world's largest producer of cloves supplying more than 90 per cent of the global crop. This it managed on the back of the intensive use of slave labour, which was used to handpick the flower buds.

In this fashion the crop jumped westwards across the Indian Ocean, with each island stepping-stone reducing the amount of genetic diversity present until it finally reached the east coast of Africa. Every link in the chain involved planting new trees from seeds and since cloves are highly self-fertile, each step resulted in more and more inbred offspring, to an extent not even matched by the best of European royal families! Even within the original populations in the Moluccas the lack of variation in the clove crop was documented as early as the seventeenth century. Such high levels of genetic uniformity present the clove breeder with severe problems. What hope is there of finding a tree with resistance to the ominously named 'sudden death' disease from among this ocean of sameness? Probably the only way to prevent future generation of cloves in Zanzibar from falling victim to this disease, which rapidly kills trees on the point of reaching maturity, is to once again cross the Indian Ocean to the Moluccas. This time the quest is not for the origin of the clove, but is a search for its wild ancestors and the Holy Grail of resistance to sudden death.

Fig. 8.1 For centuries colonial powers struggled to control the world supply of cloves, with disastrous consequences for its genetic diversity.

Rubber's busting bubble

Although not as common as it once was, one tropical tree crop seems to have found itself a new habitat in office foyers and sitting rooms across the land. However, the pot-plant with the large glossy dark green leaves that is widely known as the rubber plant is a fraud, a charlatan. It is not a rubber tree at all, it is a sort of fig. This confusion probably arose because when damaged the rubber plant of the large plastic pot drips milky white sap, just like a real rubber tree. But then there are many unrelated plants that contain latex and indeed at various times attempts have been made to harvest rubber from some of these, including the humble dandelion.

The title 'rubber tree' truly belongs to an evergreen tree, which grows wild in the Amazonian rainforests of Brazil and Peru, a fact that is reflected in its Latin name *Hevea braziliensis*. It is a member of the Spurge family, which includes another latex producing pot-plant, the Poinsettia. The rubber tree probably started its long association with mankind as a food plant. Its cooked seeds are still regularly eaten by some native South American Indians, although it is usually only consumed at times of famine. Well before the first Europeans reached South America the indigenous people were collecting latex from the rubber trees and producing an elastic substance, which they used to seal various bindings. No one is really sure when rubber first reached Europe, and initially rubber products were only considered as curious novelties, as indeed many still are. This is because items made of crude natural rubber, crumble after a short while.

In 1791, an English inventor called Samuel Peal patented a method of producing waterproof clothing by treating it with rubber and turpentine. This represented the first commercial application of rubber, the birth of the rubber suit and with it a new fetish. However, it was the discovery, in 1839 by the American Charles Goodyear, that cooking rubber with sulfur increases both its strength and elasticity, that really led to the rubber boom. It was quickly realized that this 'vulcanized' rubber being impermeable to gases, electricity, and being resistant to abrasion, water, and more corrosive chemicals could have hundreds of applications, not least the pneumatic tyre. The price of rubber soared!

At the time, all of the world's rubber was produced from trees growing wild deep in the Amazonian forests. The result was that, as the nineteenth century drew to a close, the region witnessed a rubber-rush. Boom times brought incredible riches to a few rubber barons as new towns were founded along the Amazon and its tributaries. Although the streets of Iquitos in Peru are full of cars, even to this day it can only be reached by air or river, as the roads terminate a short distance from town in the jungle. In spite of these transportation difficulties no expense was spared in the construction of these towns. Local rubber baron Jules Toth imported an entire two storey metal house that had been designed by the French engineer, Gustave Eiffel (of the tower fame) for the 1889 Paris Exhibition. Downstream in Brazil, at the confluence of the Amazon and Rio Negro rivers, the city of Manaus was for a brief moment the richest place on earth. Its newly rich citizens were famous for their extravagant gestures. The gulping of vast quantities of champagne and the lighting of cigars with hundred dollar bills gave the town a reputation for decadence. Its merchants imported marble from Italy, iron pillars from England and polished wood from France to build the magnificent Teatro Amazonas opera house, which played host to many of the great stars of the day when it opened in 1896. However, great riches are never amassed without someone else becoming envious. And thus it was, in 1876 that despite strict embargoes imposed by Brazil to protect its interest, the British explorer and botanist Sir Henry Wickham managed to smuggle 70,000 rubber seeds out of Brazil to the botanic gardens at Kew. Legend has it that when challenged Sir Henry replied that the seeds were destined for Queen Victoria's orchid collection. The Brazilians were not surprisingly rather displeased about this, and to this day remain reluctant to allow other nations to collect plants in their forests.

From these seeds the gardeners at Kew managed to raise about 3,000 plants, with the majority being quickly exported to Sri Lanka, Malaysia and Indonesia. It is thought that even now virtually all rubber growing in the Eastern Hemisphere is descended from those smuggled out of Brazil by Sir Henry. As rubber is a fast growing tree, reaching maturity in about five years, it did not take long for Wickham's rubber plants to become serious competition, breaking the South America monopoly of supply. By the early part of the twentieth century the short-lived Amazon rubber boom was over.

The cultivation of rubber reached its peak shortly after World War II. The disruption of supply that occurred during the war stimulated research into synthetic rubbers, the production of which is now considerably more important than the cultivation of natural rubber. However, natural rubber remains the preferred option for many products, and thus it was that the panic about AIDS during the 1980s saw a sharp rise in the price of natural rubber!

Stimulating tea

According to Chinese mythology more than 4000 years before British mythology claims that Isaac Newton was inspired by a falling apple to discover gravity, the Emperor Shen Nung sat beneath a tree and made, what many consider, an equally important discovery, by watching a falling leaf. Legend has it that in 2737 BC the Chinese Emperor, an amateur herbalist, observed leaves of a wild tea tree falling into a pot of drinking water that his servant was boiling. Enticed by the aroma of the resulting infusion, Shen Nung was moved to sample the brew, and it is said, he was so delighted by the taste, that he never again drank plain water. In Indian and Japanese tradition, the discovery of tea is ascribed to Bodhidharma, the founder of Zen Buddhism, who is believed to have kept himself awake contemplating Buddha for seven years with the aid of the stimulation provided by chewing wild tea leaves. All of which suggests that tea is probably best avoided at bedtime.

The Latin name of the tea tree, *Camellia sinensis*, gives away both its place of origin and the fact that it is a close relative of the garden Camellias. In the non-cultivated state, the tea tree can grow to 30 metres high, and its evergreen leaves were originally picked by trained monkeys (not PG chimps). However, in modern tea gardens the more regular picking of leaves ensures that the bushes rarely develop above one metre. Tea originated in sub-tropical South East Asia in the area between the Yangtze and Brahmaputra rivers, with distinct types developing in different regions. Some botanists have described these types (Assam, Cambodia, China and Irrawaddy) as different sub-species, with hundreds of different varieties occurring within them. But the situation is confused further by the division into, black, green and oolong teas. Although historically many Europeans thought that these were derived from different plants, they are in fact the products of different methods of processing the leaves. Nevertheless, it is true that generally green teas are made from the Chinese sub-species and black teas are made from the larger-leaved Assam types.

The best teas are made from small young leaves picked as the bush produces a mass of new growth. These are processed by being spread out on racks and allowed to wilt. The leaves are then rolled to break open their cells and mix the cell contents. This starts a complex chain of chemical reactions termed fermentation. Although referred to as fermentation, this process does not involve alcohol, but is similar to the reactions which cause cut apple and other plant material to brown in the presence of oxygen. During the fermentation phase the leaves develop flavour and colour as they are spread out on trays in a humid but cool environment. In the production of green teas the leaves are steamed which prevents fermentation. In contrast, black tea is fully fermented and oolong or literally black dragon tea is partly fermented. Finally the leaves are dried with hot air, or more traditionally over hot ashes, before grading by size.

The longer the fermentation period is, the stronger the flavour and the higher the caffeine concentration of the resulting tea. The amount of caffeine in tea is also affected by the variety of the tea, the age of the leaf, and critically on the method of brewing. Increasing the temperature of the water, allowing the tea to stand for longer and decreasing the size of the tea leaves all increase the amount of caffeine in the final drink. Caffeine is a complex molecule, which has a stimulating effect on the tea drinker. It was first discovered in tea in 1827 and was named theine. Later the same compound was found in coffee and named caffeine. Eventually it was realized that the two things were one and the same and the name theine was dropped. Weight for weight tea does contain more caffeine than coffee, but because less tea is used in making a cup of tea than coffee is used in making a cup of coffee, an average serving of tea contains about half the amount of caffeine of an average cup of coffee.

Over the years tea has not only stimulated the tea drinker, but its high value also helped to stimulate American independence and subsequently the opium wars between Britain and China. The early history of tea in Britain is obscure. The Dutch and French were the first in Europe to popularize the beverage. With the restoration of the monarchy under Charles II, tea drinking finally became fashionable in Britain. Working with the premise that if anything is popular it must be regulated, Charles II tried to outlaw the sale of tea from private houses in 1675. This failed to become law, but a year later the first duties were imposed on the sale of tea and licences were required to run tea-rooms. Tea taxes rose dramatically reaching a peak in the mid eighteenth century at 119 per cent. In addition, tea prices were elevated by the John Company.

Founded under Elizabeth I, the John Company held the most absolute monopoly of trade in world history. It had total rights to trade east of the Cape of Good Hope and west of Cape Horn, and was granted rights to pass and implement laws, mint money, raise arms and declare wars. Powers you would hope that modern multinationals do not even dream of. By 1777 the cost of a pound of tea was equivalent to about one-third of the average weekly wage. This stimulated mass smuggling of tea, via Holland and Scandinavia, with syndicates distributing the contraband across the country. In addition the adulteration of tea was widespread, and although outlawed in 1725, it was common to find tea mixed with, ash, elder and willow leaves, reused tea leaves and even sheep dung.

In 1773 the John Company was merged with the East India Company. Its new charter granted it a total monopoly of commerce with China and India and the right to by-pass the colonial merchants and sell tea directly in America. The result was to force many American tea importers out of business. This came on top of The Stamp Act – a purchase tax on many products and an importation tax on tea, paper and glass, both of which the British imposed on their American colonies to recoup the costs of the recent French and Indian War. These events outraged the American colonialists who were great tea drinkers and led to the famous Boston Tea Party, which in turn precipitated the American War of Independence. On the night of December 16th 1773, about 50 men dressed as Mohawk Indians (in protest against the rise in the tax on tea linked to the Indian war) boarded three of the East India Company's ships in Boston Harbour; the Dartmouth, the Beaver and the Eleanor. Once aboard they smashed open, or threw overboard, 342 chests of a choice black Chinese tea called Bohea, worth 9,650 pounds. In retaliation English troops occupied Boston City and closed the port through which most of America's tea was imported. The colonials responded by declaring the revolution and turning their affections from tea to coffee.

Finally in 1784, the Prime Minister William Pitt the Younger slashed tea tax from 119 to twelve and a half per cent, thus ending tea smuggling and encouraging free trade. The nineteenth century witnessed intense competition between the British and Americans to dominate the world trade in tea. The Americans literally led the race by designing new and faster tea clippers. The British quickly copied these more streamlined vessels because they halved the journey time of the older heavier tea wagons. Until the introduction of the steamship, American and British tea clippers vied to be the first each year to bring tea from China to the London Tea Exchange. Curiously the most famous tea clipper ever built, The Cutty Sark, only rarely carried cargoes of tea. The trade in tea was to create America's first three millionaires, J.J. Astor, S. Girard and T.H. Perkins.

Crippled by the outlays of cash required to purchase tea in China, the East India Company devised a solution. By growing poppies cheaply in the newly occupied India, it could exchange opium in China for tea. Not surprisingly the Emperor of China was not pleased, and the Opium wars broke out. Britain fought to protect its right to swap hard drugs for soft, which it did unmolested in China until 1908. It is rather ironic that the most 'civilized' of British traditions, the afternoon tea, was formerly paid for by the international drugs trade.

There's an awful lot of coffee

Not many crops have high street shops dedicated solely to their trade. But then coffee, which in monetary terms is the second most important commodity in the world market behind petroleum, is no average crop. It is a giant global industry employing more than 20 million people.

Coffee is unusual as a crop in that it is not a single species. About ten different species of tropical and sub-tropical evergreen trees are cultivated to produce coffee, although just two of these are of international importance, with the rest being grown solely for local consumption. These different species of coffee are closely related members of a widespread family of plants, which also includes species native to Britain. In temperate regions the coffee family tend to be small herbaceous plants, the best known example being cleavers or sticky willy. The family resemblance can be seen in the seeds of cleavers. Once extracted from your socks these appear like two small coffee beans, and indeed during the Second World War they were used as a coffee substitute.

About three-quarters of the world coffee crop is comprised of arabica coffee, a species of small tree that was originally domesticated in the mountains of Ethiopia. At some stage in its evolution, arabica coffee appears to have arisen from the other species of coffee by doubling the number of chromosomes in its cells and simultaneously changing its sex-life. This is not an uncommon phenomenon in the evolution of plants. Arabica coffee as a species of the mountains routinely reproduces by having sex with itself. It is a cold fact of life that sex with a partner tends to become increasingly difficult with altitude. In contrast the second important species of coffee, robusta is a plant of lowland tropical Africa. Discovered growing wild in the Congo River basin only in 1898, the plants of robusta coffee only produce coffee beans if they have been pollinated by another tree. Robusta coffee trees are larger, more vigorous, higher yielding, and more disease resistant than arabica trees. Sadly the coffee they produce is considered to be of inferior quality.

Coffee beans take about eight months to develop. Each fruit, which contains a pair of coffee beans, ripens to a bright red cherry-like appearance. These shiny red berries look just like the sort of things that parents warn their children not to eat, and as such coffee is one of those crops that you cannot help wondering - how did anyone think of using it in the first place? According to Arabic legend the consumption of coffee can be attributed to an observant goat-herder named Kaldi. One night Kaldi's goats wandered off on their own, after an exhausting search the goat-herder found his animals dancing friskily amongst a grove of wild coffee. Possessing a curious nature, it was not long before Kaldi was also dancing. His eccentric behaviour not surprisingly attracted the attention of the local imam, who also quickly acquired the coffee habit, which he discovered solved his problem of falling to sleep during prayers. From this beginning the use of coffee is said to have spread from monastery to monastery across the Arab world.

This story is consistent with other early records of coffee use, which suggest that originally the leaves and beans were chewed for medicinal reasons rather than being imbibed for culinary ones. This is further supported by the use of coffee as a stimulant. Formerly herbalists recommended giving coffee to persons suffering from poisoning. In extreme cases, such as the victim of a snakebite, it was recommended that the coffee be administered by injection directly into the rectum. This practice gives an entirely new meaning to the question – 'how do you take your coffee?'

The use of coffee as a beverage developed in Arabia during the fifteenth century. The Ottoman Turks introduced it to Constantinople in 1453 where it became considered such an essential part of everyday life that women were legally entitled to divorce husbands who failed to provide an adequate daily quota of coffee. From Constantinople the drink spread into Western Europe. Although the papal advisors initially tried to ban the product because of its infidel connections, Pope Clement VIII appears to have been quick to appreciate the fund raising potential of the church coffee-morning and baptized coffee, making it an acceptable Christian beverage.

The Arabs jealously guarded their control of coffee production, allowing only roasted beans to be exported to Christian Western Europe. However, eventually a Muslim pilgrim from India called Baba Budan breached the embargo. Legend has it that he kept seven coffee beans safely strapped to his belly until he returned home. The descendants of these seven beans were planted all over India. This subsequently allowed the British, French and Dutch to break the Arab monopoly by smuggling coffee to their various tropical colonies. However, the globalization of coffee was not all plain sailing. In 1723 a French naval officer, Gabriel Mathieu de Clieu failed to obtain the approval of the authorities in Paris to export coffee to the French Caribbean Island of Martinique. Undaunted, he stole some plants and smuggled them aboard ship. During the voyage he reported being attacked by a Dutch spy, who was intent on sabotaging the foundling French coffee industry by attempting to destroy the plants. Surviving the inevitable attacks by pirates and almost being shipwrecked during a violent storm, de Clieu's log records that he was finally forced to share his limited water rations with his coffee seedlings. Eventually all but one plant succumbed, but this survived to found a dynasty of 19 million coffee trees in Martinique within 50 years.

The factors affecting the quality of coffee are similar to those which influence wines during cultivation, factors being variety, soil type, climate and later, the stage the beans are picked at, plus the processing and roasting of the beans are important. Professional coffee tasters are like wine connoisseurs in having a language all of their own – pretentious twaddle! All that actually matters is whether or not you actually like the taste. There is a degree of brinkmanship in the growing of fine flavoured coffee. The best quality coffees are produced from arabica plants grown at an altitude of between 1300-2000 metres and picked by hand. Arabica coffees cultivated at altitude are referred to as high grown milds, and are consumed almost exclusively as speciality coffees. The higher the altitude the plant is grown at, the slower the berries mature, producing a smaller denser bean containing less moisture and more flavour. The risk associated with growing fine flavoured high altitude coffee is that of an occasional frost. A single cold period can be enough to damage the trees and retard growth for several years.

Today's advertisements for coffee frequently portray an image of romance and seduction. The line 'Would you like to come in for a coffee?' is usually loaded. The connection between coffee and the passionate liaison is however, nothing new. In 1727 the Emperor of Brazil was anxious that his country should break into the lucrative market in coffee. The colonial powers of France and Holland were equally keen on guarding their self-interests. Coffee plantations throughout the Guianas were heavily defended and thus it was that Lieutenant Colonel Francisco de Melo Palheta was sent under the cover of arbitrating in a border dispute between the French and Dutch colonies in Guiana, to purloin some coffee plants and return with them to Brazil. Legend has it that although he was successful in resolving the local political situation, he had greater difficulties with his main task. Undaunted young Francisco resorted to using his deadly charm and suave sophistication on the wife of the Governor of French Guiana and a secret alliance of the under-cover variety resulted. As the lady said her fond farewells to her departing lover, she presented him with a bouquet of flowers, which contained sprigs of coffee in berry. From this romantic gesture the billion-dollar Brazilian coffee industry was born. A tale of seduction to match even the steamiest of coffee adverts!

Mulberry smuggling monks

One explanation of the nursery rhyme 'here we go around the mulberry bush, on a cold and frosty morning' is that such a tree was said to grow in the exercise yard at Wakefield prison. The provision of such delicious fresh fruit for convicts appears like an act of generosity, rather out of character for the gaolers of old. Furthermore, mulberries have an ancient association with emperors and kings and not with the criminal classes.

There are three main species of mulberry commonly in cultivation: the black, the white and the red. The colour refers to the foliage as much as the fruit. All three are short trees not bushes, which produce separate clusters of either entirely male or female flowers. Occasionally individual trees may produce only flowers of a single sex. The white mulberry is native to China where its leaves have been harvested to feed silkworms for more than 4000 years. The black mulberry was originally from Western Asia, but has been grown for its fruit in Europe since pre Roman times. The red mulberry or American mulberry is a native of the eastern states of the US and is a relative newcomer to cultivation.

The history of the mulberry in China is interwoven with that of silk. The empress Si-Ling-Chi is credited with its invention in 2640 BC. After picking cocoons from a mulberry tree in the palace garden, she unravelled the stands and then spun them to make a robe for the emperor. For more than 2000 years the Chinese kept the secret of silk production under pain of death!

In Europe fanciful ideas were proposed to explain the origin of the 'cloth of kings'. It was thought to be extracted from inside the bodies of exploding spiders, or spun from soil. Other theories claimed that it was produced from petals or from hairy leaves. Eventually of course the truth did emerge. According to tradition the secret of silkworm farming was stolen by the Japanese who established their own industry with mulberries, moth eggs and four Chinese girls all purloined from the mainland. The Indian approach was rather more enlightened. It is said that stems of mulberry and silkworms were carried to the sub-continent in the headdress of a Chinese princess on route to her wedding to an Indian prince. The legend of its introduction into Europe is equally fanciful. Justinian the sixth century Emperor of Constantinople entrusted two monks with the task. After many years of toil the two ecclesiastical secret agents were successful in smuggling both mulberry cuttings and silkworm eggs out of China inside hollowed out walking sticks.

By the seventeenth century James I had become so anxious about the cost of silk imports, that he famously issued an edict encouraging the cultivation of mulberry trees and silkworm farming. Tens of thousands of trees were sold at three farthings per plant, or six shillings per five score. For a time a single tree could be rented, at a rate of a pound per year. Many of these now very old trees survive to this day, with hardly a stately home in England not claiming to have possessed one at some time. The king himself purchased a four acre plot near to his palace at Westminster on which to plant mulberries. The site, which cost 935 pounds, is now occupied by the gardens of Buckingham Palace. The price included the cost of levelling the site, erecting perimeter walling and planting the mulberries. Unfortunately the king appears to have been ill-advised because, according to legend, he was told that the white mulberry would not thrive in England. So across the land black mulberries were planted, but these are not favoured by silkworms and thus the attempt to produce silk in Britain failed.

Another legend claims that in 1609 Shakespeare obtained a mulberry tree from the king's Westminster garden, which he then planted in his own grounds in Stratford-on-Avon. The tree thrived there until 1752 when the then owner of the house, a Reverend Gastrell, chopped it down to discourage tourists. The timber from Shakespeare's tree appears to have acquired some of the properties associated with that taken from the 'true cross', in that hundreds of objects now claim to be made from it. In addition the botanic gardens at Kew claimed to have a descendant of the same tree.

To this day mulberry silk is regarded as being of the finest quality and it is still smuggled out of China. Indian newspapers complain about the negative impact that this illicit trade has on local markets. Prices are also influenced by increased demand for sarees prior to the marriage seasons of November to January and March to May. So silk farming might be a sound investment after all, but it's probably wise to do a bit more research beforehand than the British royal family did.

Fig. 8.2 Historically the value of the mulberry tree was linked to its use in the production of silk.

Fooled by the monkey puzzle

With no more than 550 men, Cortés was able to conquer the entire Aztec Empire. The Incas fell to Pizarro and his band of 180 conquistadors and a few horses. But when Pedro de Valdivia moved against the Araucanian Indians (now called the Mapuches) of southern Chile in 1541 things were to turn out differently. By 1554 Valdivia had been captured, lashed to a tree and decapitated. Some say that his Indian executors then ripped out his heart and devoured it! Not until 1881 were the Mapuches finally forced into submission and the area opened to European settlers. Imagine therefore the trepidation with which the Scottish naval surgeon and botanist Archibald Menzies sat down to dine at a banquet with the locals in 1795. A fear made even more intense as he surreptitiously slipped a few nuts from the table into his pocket. Archibald's impoliteness was not because he preferred his food deep-fried, but because he wanted to get his hands on a crop regarded as sacred by the local Indians. A crop that was their staple food: the monkey puzzle.

The monkey puzzle or Chile pine, which is known in Latin as *Araucaria araucana*, was named in honour of the Indians of southern Chile and not after the much loved compiler of the Guardian crossword. Now well known in the gardens of British suburbia, this evergreen conifer is native to Chile and Argentina where it once dominated vast tracts of forest. It can grow to a majestic 30 metres and takes 30 or 40 years to become mature. Female trees produce cones, which are the size of a human head. After two or three years maturing, these cones crash to the ground still containing their 200 plus almond like fatty seeds. The ease of harvesting the nuts, along with their high-energy value made them appear to Archibald as a potential wonder crop of the future. It was calculated that only eighteen mature female trees grown with three accompanying males were all that was required to sustain a grown man for a year. Unfortunately it is impossible to determine the gender of the trees until they start to flower. It may therefore take more than thirty years before the sex ratio can be adjusted. Today the concept of a single species diet sounds rather monotonous, but to the Scottish surgeon, accustomed to long years at sea, it was an attractive enough prospect to risk committing a serious social blunder.

Once back on board ship Menzies carefully planted his valuable seeds. He lovingly tended his monkey puzzles during the journey home around Cape Horn. During the long years on board George Vancouver's 'Discovery', in search of the fabled Northwest Passage, Vancouver had carried Menzies around the Pacific and along the West Coast of North and

South America. The rigours of the voyage had caused the death of almost all Menzies' plants. Only his dried herbarium specimens, which were to form part of the Kew and Edinburgh Royal Botanical Garden collections, and five of his treasured monkey puzzles survived intact. Thus it was that in 1795 the monkey puzzle was introduced to Britain, not as a horticultural curiosity, but heralded as a new wonder crop, destined to feed the masses. The plants were donated to Sir Joseph Banks, one of which survived at Kew until 1892.

Sadly not only did the monkey puzzle fail to make any contribution to the British diet, it has also suffered a serious decline in its native Chile due to logging. The once extensive pine forests have gone. All that remains are two small populations, one along the coast and another on Chile's border with Argentina. International trade in the monkey puzzle is outlawed under The Convention on International Trade in Endangered Species of Wild Fauna and Flora (CITES). It is rather too late to prosecute Archibald Menzies, who risked rather more than a fine, for his dream.

Prohibition of the artichoke

According to Greek mythology there was once a beautiful young girl called Cynara who lived on the island of Zinari in the Aegean Sea. One day when the king of the gods, Zeus, was visiting his brother Poseidon, he spied the lovely mortal girl by the sea. Cynara was unperturbed by the presence of the deity. Zeus must have seen this as a green light and wasted no time in seducing the maiden. Wishing to take her home to Mt Olympus, Zeus decided to change Cynara into an immortal, and so she became a goddess and went to live with the other gods. However, Zeus neglected the new goddess in favour of his wife Hera, and Cynara soon became homesick for her mother and Zinari, so much so that she decided to sneak home for a brief visit. When Zeus discovered what she had done, he was most displeased and banished her from Mt Olympus. She returned to earth transformed into the form of the plant we know as the artichoke. To this day the globe artichoke's scientific name is *Cynara scolymus*.

Unlikely as it may seem, this unusual edible thistle has a long association with beautiful women and as recently as 1947 the young Norma Jean Baker was crowned Miss California Artichoke Queen in the Castroville Artichoke Festival and so started Marilyn Monroe's superstar career.

Ever since the time that Zeus banished Cynara from Mt Olympus, the artichoke has experienced something of a rollercoaster ride in the popularity stakes. To the ancient Greeks and Romans artichokes were regarded as a great delicacy and aphrodisiac. They were thought to be a way of ensuring male offspring. Even so this opinion was not universally held as the Roman writer Pliny the Elder regarded them as "one of the earth's monstrosities". With the fall of the Roman Empire the artichoke rather fell out of favour until the sixteenth century when the fourteen year old Catherine de Medici was married to Henry II of France. Catherine is credited with popularizing the artichoke in northern Europe, and it is said that she once ate so many that she thought she would die and was very ill with diarrhoea.

Within two hundred years artichokes were no longer regarded as a fashionable food for royalty, and dismissed by the German writer Johann Wolfgang Goethe as thistles eaten by Italian peasants. Throughout history the Italians seem to have appreciated the artichoke more than any other nation, taking this fondness with them to America. By the 1920s Italian-Americans in New York represented a significant market for the crop, so much so that the Mafia turned its attentions to supplying the trade. Mafia member Ciro Terranova became known as the "Artichoke King", monopolizing the market with a reign of terror known as the artichoke wars. He destroyed his competitors' crops, hacking them to pieces with machetes. That is the artichokes not the competitors!

By 1935 the Mayor of New York, Fiorello LaGuardia had had enough and declared the situation to be a serious and threatening emergency, and so he banned the sale, display and even possession of artichokes. Unlike the prohibition of alcohol (which lasted 13 years and had ended two years earlier), prohibition of the artichoke was short lived, possibly because LaGuardia himself was rather partial to them. The ban was lifted after only a week. However, this was enough to see them plummet in price and break the Mafia's stranglehold over supply. Perhaps sadly, the ending of prohibition was too rapid to see the establishment of illegal underground dens dedicated to the consumption of artichokes or speakeasies glamorizing the image of the fallen goddess Cynara. Now of course almost every pizza restaurant openly and legally offers Quattro Stagioni (Four Seasons) pizza, with artichokes representing spring, olives for summer, mushrooms signifying autumn and finally ham for winter.

Unlike the globe artichoke, which is a glorified thistle, the distantly related Jerusalem artichoke is a sunflower. The name Jerusalem is rather misleading as these potato-like edible tubers are of North American origin, being originally domesticated along the Ohio and Mississippi rivers. As with the globe artichoke, Jerusalem artichokes have a mixed reputation. Although they are regarded as a delicacy by many, the fact that the tubers store inulin rather than starch is a mixed blessing. On digestion inulin is broken down to fructose rather than glucose, which is good news for diabetics. Unfortunately, in many individuals inulin digestion results in the production of copious amounts of flatulence and this has caused a renaming of the plant to the fartichoke!

Fig. 8.3 The Jerusalem artichoke is not as it's name suggests from the Holy Land. It is one of very few crops to have originated in North America.

Pineapple snobbery

You might think that the shape of the pineapple is so distinctive that it could not easily be mistaken for anything else. However, there are images on murals in the Roman ruined city of Pompeii and pottery models found in ancient Egyptians tombs that have been interpreted by Thor Heyerdahl as pineapples, and as such as evidence of pre-Columbian transatlantic trade. Others remain to be convinced!

The distinctive shape of the pineapple has been employed as a symbol of decadent luxury since it was first introduced into Europe from South America in the sixteenth century. In 1661 Charles II was painted being presented with a rather odd looking imported fruit. The use of the pineapple as a status symbol by the aristocracy reached its height during the eighteenth and nineteenth centuries, when armies of gardening staff lovingly tended pineapple plants within glasshouses, heated by coal fires or piles of rotting manure. Their task was to produce ever larger fruit to crown lavish tabletop displays of produce. National pride was at stake, and in 1817 the British ambassador in Paris was able to humiliate the French by insisting that it was impossible to obtain respectable fruit for a banquet without sending his diplomatic coach to London. In July 1821 Lord Cawdaw's gardener produced a massive ten pound eight ounce British grown fruit. In the United States it was possible to rent imported fruit for an evening's function, only the extremely wealthy actually purchased a pineapple to eat. However, all of these were overshadowed by the fourth Earl of Dunmore who erected a pineapple shaped folly, which still stands on his estate in Stirlingshire.

The pineapple is an unusual member of a tropical family of plants called the Bromeliads in that it is ground dwelling. Almost all of the rest of the family are to be found growing perched high in the tops of trees in the tropical forests of South America. This strategy appears to evolve in dense forest, where insufficient light reaches ground level to sustain plant growth. The pineapple is also unusual in that it differs from most crops in the evolution of its sexual preferences. It differs from all its close relatives in being unable to fertilize itself. To produce seed a pineapple must have a partner since the plant has the ability to recognize its own pollen grains and prevent them from fertilizing its own ovules. Typically during the domestication process of most crops individuals are selected which do not require partners other than themselves to ensure successful fruit production, because 'self-sufficient' individuals tend to be more efficient at producing fruit. The pineapple is an exception to this rule because the cultivated fruit, although potentially perfectly fertile, will also develop in the absence of seed, and indeed fruit containing seeds are undesirable. Thus by selecting plants which are unable to set seeds without a partner and by growing these in isolation or in blocks of genetically identical plants (by vegetative means) it ensures that only desirable seed-free fruit are produced.

In its native South America humming birds may pollinate the 100 to 200 flowers per pineapple, resulting in between 2000 and 3000 five millimetre long seeds being produced per fruit. Thus to protect the virginity of its pineapples and avoid unwanted seeds within its valuable commercial crop the US state of Hawaii took the unusual step of outlawing all humming birds, declaring them undesirable aliens.

Although wild pineapples have been reported from several places including Brazil, Venezuela and Trinidad, it is considered that these populations are most likely the descendants of abandoned cultivated plants rather than being truly wild. The domestication of the pineapple pre-dates the arrival of the Europeans, but exactly where this occurred is uncertain. The first European contact with the fruit was on the Caribbean Island of Guadeloupe, where Columbus' second voyage to the New World landed on the fourth of November 1493. Although Columbus' own log of the trip did not survive, crewmember Michele de Cuneo recorded finding fruit in the shape of a pinecone, but twice as big and excellent and wholesome. The same account goes on to describe another local delicacy in the form of two castrated boys that were being fattened for the pot. In the Caribbean pineapples are often eaten with a pinch of salt, and this may also be the best way of taking such stories of cannibalism, which were often used to justify the barbaric treatment of the native people.

In 1874 in the Azores it was discovered that wood smoke could be used to induce flowering in pineapples. This enabled fruiting to be synchronized and thus improved the efficiency of harvesting. Later when it was realized that it was ethylene in the smoke, which was the active ingredient, growers adopted the practice of putting calcium carbide (which reacts violently with water to produce acetylene, which has a similar ripening effect to ethylene) into the crown of each plant. However, too much carbide, a quick downpour and a spark, and bang - your fruit may explode. Furthermore, calcium carbide frequently contained arsenic as a contaminant, so today pineapples tend to be ripened by the use of ethylene gas.

Pineapple cultivation is not without other risks. The fruit contains bromelain, a mixture of protein digesting enzymes, which have been extracted as a meat tenderizer. These enzymes regularly digest away the fingertips of pineapple workers, but then that is a small price to pay for such a valuable status symbol.

The list of crops that could have been included in this chapter seems almost endless. This fact illustrates quite clearly that the economic value of crops has significantly influenced their domestication. Driven by human greed for the wealth associated with controlling the supply of important crops, plant materials have been stolen, smuggled, destroyed and replanted time and time again. It will be argued in the next chapter that preserving the economic status quo has limited our vision in terms of domesticating new crops. Sugarcane is a rare exception, where trying to break the monopoly of supply, directly resulted in the domestication of an alternative.

References

Most significant new sources in the order they were utilized in this chapter.

Bligh, W. (2008) *The Bounty mutiny: Captain William Bligh's first-hand account of the last voyage of HMS Bounty.* Florida: Red and Black Publishers.

Moitt, B. (ed) (2004) *Sugar, slavery, and society: perspectives on the Caribbean, India, the Mascarenes, and the United States.* Gainesville, Florida: University Press of Florida.

Tully, J. (2011) *The Devils milk: a social history of rubber.* New York: Monthly Review Press.

Willson, K.C. (1999) *Coffee, cocoa and tea.* (Crop production science in horticulture series, number 8). Wallingford, UK: CABI Publishing.

Roach, F.A. (1985) *Cultivated fruits of Britain: their origin and history.* Oxford, UK: Basil Blackwell Publisher Ltd.

Aagesen, D.L. (1998) Indigenous resource rights and conservation of the monkey-puzzle tree (Araucaria araucana, Araucariaceae): a case study from southern Chile. *Economic Botany*, 52: 146-160.

Dash, M. (2010) *The first family: terror, extortion and the birth of the American Mafia.* New York: Ballantine Books / Pocket Books.

Collins, J.L. (1961) *The pineapple: botany, cultivation and utilization.* London: Leonard Hill [Books] Limited.

9

Fifty shades of green

This final chapter identifies the fact that we appear to have preferentially domesticated plants from highly nutrient rich habitats. Neither this observation nor the role of pollination strategy had previously been considered to be important in the history of crop domestication. Earlier attempts to explain why we rely on so few crop species have argued that the limiting factor has been the availability of suitable plants. Here I conclude by proposing that what limits the number of species that we currently grow and consume, is our own imaginations, prejudices, traditions and vested interests. If this is true, in the future we may enjoy a whole myriad of new fruits and vegetables that are better for our health, and less demanding of the world's limited resources.

It is frequently but apocryphally claimed that Eskimos have 50 words to describe snow. Closer to reality, but almost never quoted is the observation that there are 45 words for shades of green in the Icelandic language. In fact in most languages there are many more words to differentiate shades of green than there are for any other colour. This is because we live on a planet dominated by the colour green, where the forces of natural selection have equipped our species with eyes that are particularly sensitive to light in the green sector of the spectrum. We have evolved as botanists with acute abilities to differentiate plant species. Intensively managed agricultural fields are a brighter green than the wider countryside and when offered photographs of landscapes, people prefer the more intensely green images. There are good biological reasons behind this bias. Our crop plants tend to be particularly bright green; this is because it is indicative that they are of elevated nutritional value.

It is unlikely to be coincidence that many of the wild relatives of our crops are found along coasts and in floodplains. These habitats are naturally highly fertile, as they are regularly supplied with nutrients courtesy of defecating seabirds or inundation by rich sediments. Wild cabbages, carrots, wild-beets, asparagus, peas, various members of the spinach family and kale all have maritime distributions, while many cereals are at home in fertile river valleys. Modern agriculture replicates these rich soil conditions by the addition of copious amounts of artificial fertilizers. Previously this occurred by mining reserves of fossilized seabird droppings, quite literally mimicking the maritime habitat. As these precious resources have been depleted, they have been replaced by mineral rich rocks and artificial nitrogen. The first of these is also a limited resource and the second requires lots of energy in its manufacture and many question the long-term sustainability of this form of farming. But for now this degree of pandering enables us to maintain the lush green appearance and highly nutritious status of our crops.

The association between our crops and natural but atypically high soil fertility is responsible for them differing from the majority of undomesticated plants, which are associated with lower nutrient conditions. Living within and around the roots of many plants are symbiotic fungi. Recent analysis reveals that individual plants may have several hundred species of fungi within their root systems. As yet we still don't know what some of these fungi are doing, but many are mycorrhizal symbionts. Since the filamentous growth of fungi (known as hyphae) is much finer than plant roots, they are able to be in much closer, more intimate contact with the soil. Thus fungi are more efficient than plants at extracting available nutrients. In exchange for transferring some of these minerals to their host plant, these subterranean fungi receive sugars, which are manufactured by photosynthesis. In modern agricultural systems these relationships break down. When large volumes of artificial nutrients are added to the soil, nitrogen, phosphorus and potassium are all available to plants in excess, and they no longer need to trade their precious sugars to maintain their growth rates. As a consequence, under these conditions soil fungal diversity rapidly declines.

Finally, we are starting to understand what qualifications plants require on their CVs if they are to successfully apply to become crops. These include: being highly nutritious either by having a valuable store of carbohydrates, or having elevated mineral status by virtue of being nitrogen fixing such as the legumes, or being associated with highly fertile soils. Another important skill set that qualifies some plants to be employed as crops is the ability to store easily. Having grains, seeds, or fruit with a long shelf life is a real plus point. An alternative route into the pantheon of the farmed is a strong personality. Or more accurately, to contain a range of phytochemicals that mean you taste or smell wonderful, fight bacteria, or have mind-altering powers! Conversely, and perhaps surprisingly, being toxic does not appear to prevent plants from being successfully domesticated. Although part of the process of domestication frequently includes selecting for reduced toxicity, many important crop plants still contain potentially lethal compounds.

As a potential crop, in addition to having one of these attributes, your prospective employer will ask politically incorrect questions about your sexuality. Most crop species tend to be primarily pollinated by generalist insects. This enables them to be successfully grown in regions where different groups of insects are found. However, the bulk of what we consume is actually derived from cereals, which are wind pollinated. As with generalist insect pollination, these wind pollinated species can be grown almost anywhere, because there is always something in the breeze to tickle their fancy and enable them to set seeds. It appears therefore that the selection criterion for crop domestication includes more characteristics listed as desirable rather than essential ones. It also seems that in the absence of an understanding of genetics, serendipity has played a significant part in the process. Even so, the number of successful candidates remains remarkably low.

In science, true genius is often a matter of identifying significant questions rather than necessarily successfully answering them. The difficulty is that these questions are often so obvious that they go completely unnoticed. Professor of Geography at the University of California and all round polymath, Jared Diamond, was the first to recognize the fact that agriculture is reliant on few species of both plants and animals and that this demands explanation. Professor Diamond regards domestication as the most significant event in human history and that understanding why farmers exploit some species and not others helps explain why the modern world is dominated by Eurasian cultures. According to Jared Diamond, it is limitations within the plants themselves that has restricted the number of crops that are available. To support this theory, he identifies six independent pieces of evidence:

1. Historically crops introduced from Europe and Asia into other areas, have been rapidly assimilated suggesting that there is a limited supply of species that make ideal crops.

2. For most crops the process of domestication occurred a long time ago and very rapidly, which implies that humans very quickly identified the plants with the potential to make good crops.

3. Many important crops have been domesticated independently more than once. This is evidence that these plants really are special to have been chosen over others on several occasions.

4. There are very few modern crops. This indicates that we have exhausted the supply of plants that are suitable for domestication.

5. Pre-agriculture humans utilized significantly more species of plants than we do today. This observation implies that we originally 'screened' lots of possible alternatives before settling on our current limited selection.

6. There are many characteristics that have been identified in wild plants that limit their utility as potential crops. Finally what makes most plants unsuitable for being crops is that they are difficult to domesticate because they have undesirable properties.

Jared Diamond's list is rather different, longer and more involved than the constraints that have been identified here. So let's consider each of his lines of evidence in turn.

1. As we have seen it is generally true that new crops introduced into a region are relatively rapidly accepted by the local human population. However, there has often been a lag period as either the plant or the human have taken a while to adapt. Implicit in Diamond's focus on the now global dominance of Eurasian crops is the assumption that they are in some way superior. The same argument could also be applied to other elements of these cultures. But it is perhaps less easy to argue that the now universal attire of the baseball cap, t-shirt or suit and tie, are fundamentally better designs than the traditional garments they replaced. In fact endemic crops seem to have been more successful in avoiding being supplanted by incomers than have other elements of local tradition. This argument also applies to agricultural systems as well as crops. Time and time again, traditional diverse farming systems based on rotations and mixed plantings have been replaced by monocultures. This is not to say monocultures are necessarily superior. Ecological theory increasingly suggests that multi-species systems are more resilient to vagaries of the weather, are less prone to pests and diseases, and are potentially higher yielding than monocultures. It seems entirely probable that the reason Eurasian crops have become globally important, is that they have been promoted by dominant cultures and colonial powers, rather than them being genuinely better crops.

2. The argument that most crops were domesticated rapidly a long time ago seems a gross simplification. Domestication frequently appears to be more of a gradual process, rather than a specific event. Some crops, like the cabbage, have slipped in and out of cultivation. In some cases

gene-flow between crops and wild populations makes defining cultivated and wild plants rather arbitrary. In chapter two we discovered that although cacao has been domesticated for several thousand years, most of the world's chocolate is derived from plants that can trace their ancestry to wild trees growing in the Amazon, within a generation or two. In the kiwifruit, domestication hardly seems to have started, and wild and domesticated types merge and split at will, making species boundaries seem rather arbitrary.

3.   It is difficult to argue that a number of important crops have not been domesticated independently on more than one occasion. We have seen this in both the cabbage and several cucurbits, though it is not a widespread phenomenon. More compelling is the observation documented in chapter seven, that four main plant families have been repeatedly domesticated by the entire world's developing civilizations. However, what these facts demonstrate is that certain plants or plant families are especially suited to being domesticated, rather than that the majority of species are impossible to convert into crops.

4.   Jared Diamond's claim that there are no significant modern domestication events is perhaps the easiest to counter and the most informative in terms of understanding the process of domestication. The two most common plants in the UK are now perennial ryegrass and white clover. These two species are now widely grown across most of the world's temperate grasslands. However, less than one hundred years ago these forage species were significantly less common, but were truly wild plants. Even today they remain widespread in many ancient grasslands, but typically only form a minority component of the sward. Over the last century they have been subjected to intense plant breeding activity and now most fertile agricultural grasslands, lawns and sports fields are regularly re-sown with varieties of these species that have a shelf life of a few years. This rapid rate of turnover of cultivars is a measure of the pace of domestication of these species. These changes have gone un-noticed by most people as they stroll across the park grassland. However, those in the agricultural industry know that there are ryegrass and clover varieties available that are remarkably different from their wild ancestors. Testament to this is the fact that farmers are regularly prepared to invest the extra cost of seed to acquire the increased yields, stress tolerance, elevated sugar concentration or particular growth form offered by the latest variety.

Given that such important domestication events have occurred in recent times, it would seem sensible that we review the factors involved because these events are well documented. Rather than struggle to infer what happened from a 7000 year old, proto-cucumber that left little evidence of its existence. The publications of early grass and forage breeders reveal that they did not instantly identify ryegrass and clover as the main focus for their activities. Before they settled on this combination of grass and legume they considered a wide range of potential alternatives. Surprisingly this included ribwort plantain, which is now regarded as a weed. What appears important is not that the breeders initially screened a number of alternative species, but how rapidly these were dropped to concentrate their efforts on the 'chosen ones'. There were a number of reasons for this. Expertise in the agronomy and genetics of the selected species was rapidly acquired, while the memory that other species were even considered was equally as quickly lost. The main factor driving the concentration on a few species appears to have parallels to the process of evolution.

In 1932 the great American evolutionary biologist Sewall Wright proposed the 'shifting balance theory'. According to this theory, it is difficult for species to shift from one evolutionary solution to an alternative but better solution, unless all the steps in the process are enhancements over the previous ones. Sewall Wright used the metaphor of a hilly landscape to help explain this process. Once species arrive at the summit of a small mound it is difficult for them to un-evolve and descend this small hill before they can access any higher (fitter) peak. A very similar process is likely to work with crop domestication. Once a wild ancestor species has been selected for domestication, it may rapidly improve making it better than the non-domesticated alternatives. That is not to say that the non-selected wild plants would not eventually make more successful crops. But there is a real risk involved for a farmer abandoning the work of previous generations and gambling on an alternative, when all that is certain is that it is currently less good. Add to this the natural conservatism of farmers and the sometimes enormous economic self-interest of those promoting the status quo described in the last chapter, and you may start to believe that the failure to domesticate more species of crops lies within humans rather than within the plants themselves.

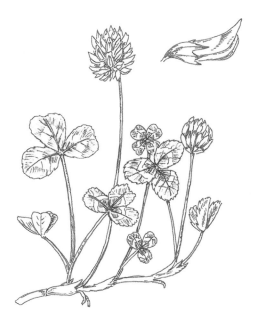

Fig. 9.1 Although white clover is a newcomer to domestication, it is now a key part of a classic grass-legume combination.

5.   The penultimate piece of evidence proposed by Jared Diamond to support his theory that plant suitability is the limiting factor in the domestication is the fact that pre-agricultural cultures harvested and ate a wider range of plants than modern people do. While this was undoubtedly the case, this observation does not account for the large number of plants that modern humans cultivate as spices, herbs, biofuels, fibres, ornamentals and for timber, medicines and recreational drugs. Virtually all of these are now derived from managed populations of genetic variants that are different from their wild relatives. In reality given that we now live in a global market place, all of us probably exploit a wider range of plants than did any of our distant forebears. Rather than plants being limited in their ability to be domesticated, we have seen that there is

apparently an almost infinite number of ways in which humans have exploited them.

6.   Finally, Jared Diamond identifies a number of characteristics that have prevented some potential crops from being exploited. This argument is similar to the one described earlier. The example Diamond cites is the comparison between domesticated almonds and unexploited toxic acorns. This has been dealt with in chapter three. In this book I have argued that plant reproductive habits have been highly influential in determining which species have been cultivated or not. This explains why the orchids, which are the most diverse of all plant families, contribute so little to our diet. Similar reasons were used to explain why temperate fruit trees are insect pollinated, while most deciduous forests are dominated by wind-pollinated tree species. Having said that, the sexual habits of plants are generally rather liberal and open to modification. The ability to self-pollinate has frequently been selected for during the process of domestication. Thus, while the characteristics of some plants appear to make them more amenable to domestication than others, the fact that our crop plants are so diverse indicates that almost any species has the potential to become a crop. This is further supported by the observation that many important food plants, such as cassava, potatoes and even wheat, still contain high levels of toxic chemicals. If being poisonous does not stop a plant from being an important food crop – then what else can?

The nub of Diamond's argument is that there are so few domesticated species because of limitations within the plants themselves. The rejected uncultivated hordes are toxic or at least unpalatable. The polygenenic control of their distastefulness means that selecting for edible variants was challenging. Thus our distant ancestors rejected these unsuitable candidates generations ago. To enforce this point Diamond asserts that in contrast the favoured few were easy to identify. Thus they were identified on several occasions or spread around the world by virtue of their desirable rare characteristics. In the previous chapters I have proposed that sometimes the sexual practices of plants have been influential in limiting the utility of some species, such as the orchids, that otherwise appear to be ideal candidates for cultivation. However, and perhaps surprisingly, chapter four illustrates that being highly toxic does not appear to disqualify a plant from being successfully domesticated. Indeed almost the exact opposite appears to be true. We have found that many of our most important staple foods are likely to contain an armoury of toxic chemicals designed to protect their valuable energy rich storage organs.

Having ruled out being toxic as limiting a species' potential to be domesticated, I have proposed that the most significant factor responsible for our narrow choice of food plants is ourselves. Our natural conservatism with trying new foods means that once we have become familiar with eating certain plants, we are unlikely to experiment with wild alternatives, except in times of famine. The rapid increases in yield, or palatability associated with the first few generations of selection is likely to result in it being risky to just throw away these advances and revert to an undomesticated species. Within living memory documented domestication events illustrate that the knowledge of considering alternative species can be very quickly lost, while familiarity with the cultivated forms is equally rapidly assimilated. On top of this the economic self-interest resulting from growing or trading in crops of commercial importance (see chapter eight) means that throughout history, powerful forces have be mobilized against those who threatened the status quo by trying to develop alternative crops. All of these factors have combined to prevent us from domesticating the majority of wild species. If we compare the modern carrot to its small fibrous ancestor, only the most vivid of imaginations could envisage what could have been produced if other lines of domestication had been pursued.

Ultimately we need to ask, does it actually matter why we consume so few plant species? Is it important if the failing lies within the plants as Jared Diamond has argued, or is it the result of innate human conservatism and an unwillingness to start over again? If the latter alternative is correct, then a whole new world of possibilities may be available if we can liberate our imaginations. In fact I have suggested earlier in this chapter that it may become a necessity to try to develop new crops that can thrive in lower nutrient conditions than those currently associated with modern intensive agriculture. Plant breeders are already starting to explore such possibilities. As our planet experiences rising temperatures related to elevated carbon dioxide levels, crop geneticists are starting to develop entirely new crops, such as elephant grass and even seaweeds, that have enhanced powers of photosynthesis and are able to fix more carbon than ever before. In future, as the natural deposits of phosphate and potash are depleted, we may need to consider domesticating plants that are mycorrhizal and able to thrive in low nutrient conditions, to replace our current 'nutrient hungry' generation of crops.

Fig. 9.2 Relatively recently perennial ryegrass was a fairly obscure grass of fertile flood meadows. Now vast areas of temperate agricultural land are dominated by monocultures of this species.

Historically crop domestication occurred without an understanding of the science of genetics. New molecular methods that enable entire genomes to be sequenced with ease mean that potentially entirely novel crops can now be produced at a rate faster than ever before. The only limitation remains our human reluctance to embrace such change. The study of the history of crop domestication has shown that for millennia the process has been anything but natural. Many of our traditional crops are artificial hybrids of uncertain parentage, containing thousands of genes from several unrelated species. In spite of this, many people are suspicious of plants that contain one or two genes from a well-known origin. These genetically modified crops are labelled as un-natural and shunned by those with as much scientific understanding as the ancient farmers that gave us poisonous potatoes, cyanide stuffed cassava and gluten rich wheat. In reality for most organisms on earth, for most of the history of life on our planet, the only way genetic mixing has occurred is by genes jumping between species. The species concerned are bacteria and viruses, which regularly pass genes horizontally between each other. In fact, viruses frequently contribute in moving genes between totally unrelated higher organisms. Thus, we all contain such genes! We are all genetically modified! What might more reasonably be considered as un-natural is sexual reproduction. Sex is a relatively recent evolutionary innovation, practiced by a minority of species. It turns out that we are all being genetically modified all the time. But then it would be really radical to suggest that human prejudice against individuals with different sexual practices from those considered normal, could limit our ability to embrace change and to develop a better more sustainable future.

References

Most significant new sources in the order they were utilized in this chapter.

Diamond, J. (2002) Evolution, consequences and future of plant and animal domestication. *Nature*, 418: 700-707.
Diamond, J.M. and Ordunio, D. (2005) *Guns, germs, and steel: the fates of human societies*. London: W.W. Norton and Company Ltd.
Voisin, A. (1960) *Better grasslands sward: ecology, botany, management*. London: Crosby Lockwood and Son Ltd.

Index